W9-AMU-331

THE ILLUSTRATED GUIDE TO
COWS

How to choose them – How to keep them

B L O O M S B U R Y
LONDON • NEW DELHI • NEW YORK • SYDNEY

First published in 2014

Bloomsbury Publishing Plc, 50 Bedford Square, London WC1B 3DP
Bloomsbury USA, 175 Fifth Avenue, New York, NY 10010

www.bloomsbury.com
www.bloomsburyusa.com

Bloomsbury Publishing, London, New York, Berlin and Sydney

A CIP catalogue record for this book is available from the British Library
Library of Congress Cataloging-in-Publication Data has been applied for

Commissioning Editor: Julie Bailey
Design by Julie Dando at Fluke Art

ISBN 978-1-4081-8135-5

Printed in China by C&C Offset Printing Co Ltd.

This book is produced using paper that is made from wood grown in managed sustainable forests. It is natural, renewable and recyclable. The logging and manufacturing processes conform to the environmental regulations of the country of origin.

10 9 8 7 6 5 4 3 2 1

Contents

Introduction

Cattle are ungulates (meaning that they are hooved animals) and it is thought that they have been domesticated for around 10,000 years. All present-day cattle are descended from the prehistoric Aurochs that migrated across Asia from India, reaching Europe about 250,000 years ago. Amazingly, the last Aurochs died in Poland as recently as 1627. Aurochs were hunted by humans long before they were domesticated, as can be seen in prehistoric cave paintings in many parts of the world, the most famous being those at Lascaux, France.

Auroch cave painting

Cattle are not indigenous to the Americas but were first introduced into the Caribbean area in the early 1500s by the Spanish. Dairy cattle were introduced by British settlers in the early 1700s.

Cows were revered in Egypt – so much so that whole cows were sometimes buried with their owners. The Egyptians worshipped the cow goddess Bat, protector and mother of the pharaohs. In the Hindu religion of India, cows are considered sacred and should be treated with the same respect as one's own mother because of the milk they provide. They appear in many stories from the Hindu faith, chief among these being the tale of Shiva, who was said to ride on the back of a bull called Nandi; and Krishna, who was brought up by cowherds and named Govinda or 'protector of cows'.

The Great Chicago Fire of 1871 that destroyed much of Chicago has long been blamed on Mrs Kate O'Leary's cow kicking over a lamp. In 1997 the Chicago City Council, after much research, passed a resolution exonerating Mrs O'Leary and her cow.

Nandi bull

There are many other traditions relevant to cattle. The ox is one of the 12-year cycle of animals in the Chinese zodiac, the constellation of Taurus represents a bull and the Maasai of East Africa believe that all cows on Earth are the God-given property of the Maasai. The ancient Romans valued their cattle so highly that the word for money, *pecunia,* is derived from the word meaning cattle, *pecus.*

Cattle are one of the most undemanding and rewarding domestic animals to keep, being in the main healthy and temperate. They are the smallholder's staple, providing the essentials of milk and beef. There are numerous breeds to choose from, many of which are described on the following pages. Some may be unsuitable for your area, but there is sure to be one breed that will be ideal for you.

A cow was connected with the founding of Durham Cathedral in AD 995. Legend has it that monks carrying the body of St Cuthbert were led to the location by a milkmaid who had lost her dun cow. The cathedral was built upon the spot where the cow was found resting.

What to consider

Are you thinking of a house cow, just a few stores or a small herd of a pedigree breed? Do you want to milk your cows, or are you only interested in keeping them for beef? Before anything else you must consider how much land is available and how suitable the grazing will be.

Cows have no upper front teeth and graze by wrapping their tongues around grass and pulling it up by gripping it between the upper gum and bottom teeth; the grass must be long enough for them to do this. Cattle are ruminants, which means that they have a digestive system that allows them to digest food that would otherwise be indigestible, by a process of repeated regurgitation. The food, referred to as 'cud', is then re-swallowed. Cattle have one stomach but it is divided into four parts, the rumen, reticulum, masum and absomasum. The cud is chewed for up to eight hours each day and the animals spend about six hours eating.

Land

Dairy cows need richer grazing than beef cattle and are likely to be of a more delicate disposition. You need to find the breed that best suits your ground. The Highland and Galloway, for example, have been bred to make the most of poor grazing in harsh conditions, and the Jersey and Guernsey need lush grass to produce their creamy milk – there is a breed to suit every need.

A question that is often asked is how many cows can be kept per acre, but this is akin to asking how long is a piece of string. What grade is your grazing? How dry is it? This is an important question, as you must also consider if your animals will need housing in winter. On good grazing consider keeping one cow to the acre, but you could increase the number of cattle if you are able to rotate your animals and take them off the grazing in the winter. On poor-grade grassland you might need five or even ten acres per beast. The size of the animal will also be a factor – a Dexter is not going to require nearly as much feed as a Belgian Blue, for example.

Ask yourself the following:

- ❑ How much land is available?
- ❑ What is the grade of grazing?
- ❑ How well drained is the land?
- ❑ Will the animals have to winter inside?
- ❑ Are you going to breed the animals?
- ❑ Do you want to sell the beef or milk?
- ❑ Is the beef or milk just for you and your family?

Fencing

It is your responsibility to fence in your livestock rather than your neighbours' to fence it out. Although cattle are on the whole calm animals, they will always have their eye on the main chance. This applies particularly to steers, which seem to have a permanent desire to be on the other side of the fence. Fence posts and gates also make convenient scratching posts, and for this reason alone they should be sturdy. If you are lucky enough to have a stock-proof hedge this will be appreciated by livestock as shelter from wind and rain. Cattle can also be very successfully contained with electric fencing, but this must be checked daily as they will notice very quickly if the battery is flat or the electricity is disconnected.

Water

All cows drink a lot of water and must have access to ice-free tanks or streams at all times. A lactating dairy cow can drink an astonishing 150 litres (264 pints) of water a day, which is not surprising when you consider that she may be producing 18–24 litres (32–42 pints) of milk.

Storage

You will need storage space for hay, straw or silage, and vermin-proof containers for any hard feed you may wish to store. Simple metal dustbins or even old freezers can be used as containers as long as they are rat and mouse proof.

When galloping through boggy, soggy places or deep mud, cattle can run faster than horses. They have cloven hooves and their toes spread so their wide feet do not sink as deep as those of the solid-hoofed horse.

Housing

Unless your land is particularly dry you may have to consider winter housing for your cattle, otherwise the ground may become poached. Provide as large an area as possible, with plenty of space for the animals to lie down in. Ventilation is very important, and ideally one side of the barn should be open to a yard area to enable the animals to take exercise. Deep litter bedding is fine, but clean bedding must be added frequently. As a consequence of this the level of the floor will rise, and this must be taken into account when siting water troughs and so on. At least 2 m (7 ft) of trough space per animal with large horns is required, although the space can be less for animals without horns.

Elm Farm Ollie was the first cow to fly in an aircraft, doing so on 18 February 1930 as part of the International Air Exposition in St Louis, Missouri, United States. On the same trip, which covered 72 miles from Bismarck, Missouri, to St Louis, she also became the first cow to be milked in flight. This was done ostensibly to allow scientists to observe midair effects on animals, as well as for publicity purposes. A St Louis newspaper trumpeted her mission as being 'to blaze a trail for the transportation of livestock by air'.

Elm Farm Ollie was reported to have been an unusually productive Guernsey cow, requiring three milkings a day and producing 24 quarts of milk during the flight itself. Wisconsin native Elsworth W. Bunce milked her, becoming the first man to milk a cow in mid-flight. Elm Farm Ollie's milk was sealed into paper cartons, which were parachuted to spectators below. Charles Lindbergh reportedly received a glass of the milk.

Feeding

Feeding is very much a matter of experience and is also totally dependent on the type of breed you own, weather conditions, available land, time of year, and whether the animals are simply being maintained or destined for meat production. It also depends on whether they are wintered inside or outside. Whatever feeding method you employ, remember that the hungrier an animal is the more likely it will be to attempt to find its own food, particularly if some is visible over a fence. Bear in mind that an average dairy cow eats up to 40 kg (88 lb) of food a day.

The main source of food will of course be grass, and between May and November your cattle should be able to survive on this alone, depending on the quantity and quality of it. There are, however, plenty of supplementary feeds available, and in most cases you will need something to get your animals through the winter.

Hay

You can make your own hay if you have the machinery, or you can buy it direct from the field. Collecting some yourself will be considerably more economical than having it delivered, although it is said that hay 'makes' better if it has been stored in large quantities – it goes on drying for a month or two and does so better in a large pile – so it may be superior in quality if you buy it after it has been stored in a large barn.

Nowadays you may need the equipment to deal with large, round bales, and these will also be better value than small ones.

Straw

Cattle are happy to eat straw, barley straw being the most popular, but although providing bulk it does not contain the same nutritional value as hay. It is a by-product of the grain industry and as it is harvested after it has set its seed, all the nutrients have left the stalk and leaves, and gone into the seed.

Silage

Silage is cut grass that is basically fermented but palatable. It can either be kept in a large pile called a clamp and covered with plastic sheeting or – more commonly today – wrapped into large bales. If the plastic wrap gets punctured the silage will spoil, so once a bale is opened the silage must be consumed within a few days.

Haylage

This is a sort of cross between hay and silage, and is grass that has been partially dried before being contained in airtight, plastic-wrapped bales. It is high in protein and is a very good feed, but is considerably more expensive than either hay or silage.

Concentrates

You can mix your own hard feed using straights such as oats, barley or sugar beet, or buy commercially produced coarse ration. Your local feed merchant can also supply all kinds of nut, cube, pencil and cattle cake. There will be beef nuts for growing and store cattle, calf-starter nuts, beef-finishing nuts, specially formulated nuts for dairy cows and even calf museli, which is a coarse ration and sometimes more palatable than nuts.

Roots

Your animals will be delighted to be given manglewurzels (or mangolds), turnips, swedes or fodder beet.

By-products

You may be able to find alternative feeds that are by-products of various industries, such as brewers' grain, which is a by-product from the production of beer and is an excellent source of protein, or wheat middlings from a flour mill. What you can find very much depends on what is available in your area.

Vitamins and minerals

These play an important part in cattle nutrition. Inadequacy of any essential minerals results in inefficient feed conversion, decreased reproductive performance and poor immunity. They can be fed in the form of mineral licks, which are available in several forms and varieties, including salt. Another method is to administer an all-trace element bolus, which is a sort of pill that dissolves over 240 days and contains around six trace elements and three vitamins. You will need an applicator to administer the bolus.

A bull or artificial insemination

Whether or not to keep a bull is a big decision. It will probably hinge on how many animals you have – any less than ten and you should seriously consider using artificial insemination (AI).

Keeping a bull is a big responsibility. Although for a good deal of the year a bull can run with the cows, any young heifers or bull calves over six months old will have to be removed. A bull will also require housing and feeding, and although he may appear to be a pussy cat you must never forget that he is a bull and is therefore unpredictable.

Never turn your back on a bull. Bulls are on the whole aggressive – it is in their nature to be so. You must also consider the expense of acquiring a bull – as he is to be the sire to your herd, the very best you can afford should be your aim

If you use AI, you can choose the breed of your choice from the very best available. You also do not need to use the same bull for every cow. You need to find the nearest AI suppliers to you and let them know the semen you have chosen. They will get it transferred to them and store it frozen until you call to say that your cow is ready. The semen comes in straws. The disadvantage of this method is that insemination must be carried out 12–24 hours after the beginning of oestrus, and all your cows may not come into season at the same time.

The world record for the heaviest bull, a Chianina named Donetto, was an amazing 1,740 kg (3,836 lb). He was exhibited at the Arezzo show in Italy in 1955.

There is a way around this problem, however. It is known as synchronization and allows groups of cows to be prepared for breeding at a time selected in advance. A vet implants a removable hormone implant into each cow's vagina for a specific length of time. AI can then be carried out between 48 and 72 hours after the implant has been removed.

Rules and regulations

There are very strict rules and regulations relating to cattle owning as well as cattle moving, and these are in place to maintain a system of traceability, particularly in the event of a disease outbreak. There are all sorts of rules and regulations and some of these change frequently, so always check the Department for Environment, Food and Rural Affairs (Defra) website for updates.

Before you do anything else

In the UK, once you have decided that you want to keep some cattle you must by law acquire a County Parish Holding (CPH) number for the land where the cattle will be kept.

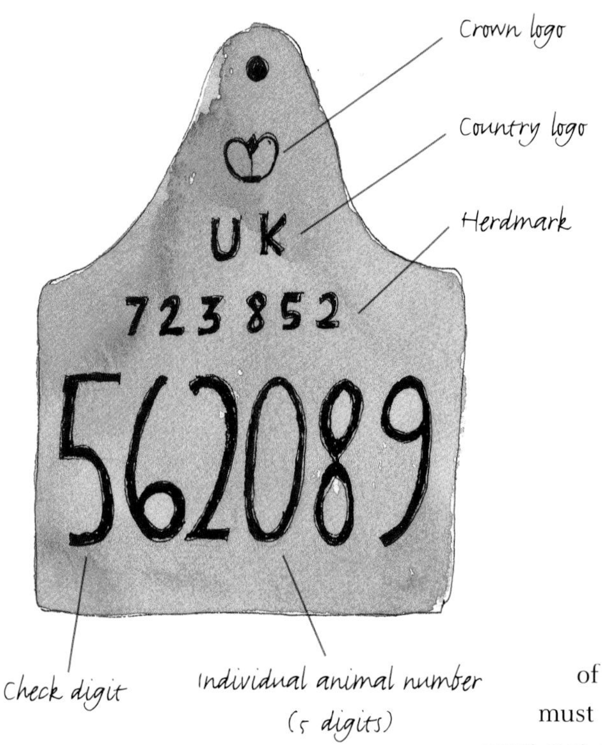

A CPH is a nine-digit number. The first two digits relate to your county, the next three relate to the parish and the last four comprise a unique number for you. You apply for a CPH from the Rural Payments Agency, and it supplies you with a personal number. If you want to keep other animals, such as sheep, pigs or chickens, you only require one number.

You must then notify your local Animal Health Office (AHO) and give it your CPH number, and it will issue a herd mark. All cattle must have two ear tags that contain the crown logo, country code, herd mark, individual animal number and check digit. Each animal also requires its own passport. When you apply for your cattle's tags the British Cattle Movement Service (BCMS) will automatically post your passport-application forms. Ear tags must be fitted within 20 days of an animal's birth, with the exception of dairy cattle, which must have at least one tag fitted within 36 hours of birth. You must not remove or replace ear tags without permission from the BCMS. If they are lost or become illegible, they must be replaced with tags bearing the same number within 28 days of discovery.

Notification of cattle movement

Whenever you move cattle on or off your holding you must follow the conditions of the general licence for the movement of cattle (available on the Defra website). Once you move cattle or other animals on to your holding, no other animals are allowed to move off it for six days – you can apply for an exemption to the standstill rule if your animals have been attending a show. All movements must legally be notified to the BCMS within three days of the event, and recorded in the movement-summary section of the passport.

The death of an animal on your holding

If an animal dies on your land the death must be reported either electronically or on the death details section of the passport. In either case the passport must be returned to the BCMS within seven days. You may not bury or burn carcasses on your holding. If the animal is more than 48 months old, you must send it to an approved sampling site so that a brainstem sample can be taken for bovine spongiform encephalopathy (BSE) testing.

Keeping on-farm records for cattle

According to European law, you must keep detailed herd records. These should contain the following for every animal:

- ❏ Ear-tag number.
- ❏ Date of birth.
- ❏ Breed and sex.
- ❏ Dam ear-tag number.
- ❏ Details of movements on and off your holding.
- ❏ Information on where an animal has been moved from and to.
- ❏ Date of death.

Animals born since 1 January 1998 must retain the same number throughout their lives. You will also need to update these records within the following deadlines:

- ❏ Thirty-six hours for movement on or off a holding.
- ❏ Seven days for the birth of a dairy animal.
- ❏ Thirty days for the birth of a non-dairy animal.
- ❏ Seven days for recording death.
- ❏ Thirty-six hours for replacing ear tags (only if the ear-tag number has been changed).

These records can be paper based or electronic.

Farm inspections

You may receive on-the-spot checks, which are referred to as Cattle Identification Inspections (CIIs). These are carried out to ensure compliance with all identification and registration requirements.

The inspector will check that:

- ❏ Your farm records show which animals are present on the holding or have been on the holding.
- ❏ Births, movements and deaths have been correctly recorded.
- ❏ All animals are correctly tagged, and match their passports.
- ❏ All passports are present and correct.
- ❏ Deadlines for identifying cattle and keeping records have been met.
- ❏ All passports for animals no longer on the holding have been passed on to the new keeper or returned to the BCMS.

Subsidies

The Single Payment Scheme (SPS), which is part of the Common Agricultural Policy, is the principal agricultural subsidies scheme in the EU. The rules as to who should apply and how change constantly, so consult the relevant website.

Acquiring stock

The obvious place to buy stock is a livestock market, and this will be fine if you just want to produce some beef for your family. However, if you wish to set up a pedigree herd you will do much better to go directly to a breeder. An enormous advantage of doing this is that the breeder's knowledge will be available to you, and you will have someone to turn to with problems.

Pedigree stock is more expensive than unregistered cross-breeds, but your calves will also command a higher price should you wish to sell them.

There are various ways to start a herd. You can acquire weaned heifer calves, but bear in mind that they will not be able to start breeding until they are 18–24 months old. You could buy an older heifer ready to breed or an in-calf heifer. You can even acquire in-calf cows, perhaps with a calf at foot.

Handling

There will be times when you need to handle your cattle – to administer a drench, perhaps, or to trim a hoof. What you need is a cattle crush, which is purpose built for the job. It can be acquired secondhand at a farm sale, or you can rig up your own by using gates – having a corral or fenced off corner of a field will make it easier to herd the cattle into a manageable area and thereby persuade them into the crush.

Halter training

If you only have one or two animals, training them to be led by a halter will be very useful – and if they are happy to be tied up you could also manage without a crush. Of course, if you start with a halter on a calf from the very beginning, there will be no problem. If, however, you have acquired an untrained cow you will need patience to train her.

One word of warning: when leading an animal never wrap the rope around your hand – even a sheep is stronger than you are and your hand could become trapped should the animal tug on the rope. Hold the rope in the palm of your hand and use two hands if necessary.

Start by taming the animal. Get it used to having its head stroked, preferably while feeding, until it is confident in your presence. When you feel that the time is right, slip the halter on and let the animal out still wearing it. You may be lucky and find that the animal will be quite happy to be led around with no problem, but it is more likely that the moment you put pressure on the rope it will pull back, and will be stronger than you are.

In the latter case the animal must discover that it is not stronger than the rope, and the only way to bring this about is to tie the rope to a purpose-built ring in a wall. For safety, tie a double loop of baler twine (it will snap if things get really out of hand) to the ring and fix the lead rope to this with a safety or quick-release knot. As soon as the animal discovers that it is restrained it will pull back with all its strength and may even throw itself on the ground, which can be alarming. Eventually, however, it will realize that not pulling makes life easier and more comfortable. You may have to go through this procedure more than once, but eventually you will have a calm and tranquil beast that will do your bidding.

Calving

There is nothing like getting a bit of hands-on experience before your first calf is born, and even more so if you will have an inexperienced heifer giving birth. Find a local farmer who is willing to let you help – it could be at the farm from which your animals came in the first place. You will need an experienced ally to calm your nerves and help decide if the vet needs to be called when the time comes.

You will need to be able to keep an eye on your cows, but obviously cannot watch them day and night. If your animals are calving outside, bring them to the nearest possible area to your barn or shed in case there are problems and you need to transfer them to somewhere that is enclosed. Never call the vet if your animals are out in a field – it would be a waste of his and your time chasing them around. If the cows are inside make sure that there is an enclosable corner.

Normal gestation in cattle is 283–284 days, but this can vary considerably on either side.

Signs of imminent calving

There are various signs which indicate that a cow is about to give birth, and it is the owner who has observed his animals who will be the most likely to spot a change in behaviour. The following are the most obvious signs:

The udder of a cow may 'bag up' several days, or even weeks, before calving is due, although some cows may bag up when calving is imminent. If yellow colostrum is seen dripping this indicates that calving will take place within 24 hours.

Cows are herd animals so if you see one going off on its own, whether in the field or a corner of the yard, this is a good indication. A cow may paw the ground, smell it, lie down and get up.

The cow may swish her tail around and you may be able to see the ligaments around the top of the tail standing out more than normal.

The vulva will expand and some mucus may be visible.

Normal presentation

A normal birth can take half an hour in a cow, or two or three hours in a first-time heifer. In a normal presentation the calf appears head first, with a nose and two hooves visible. Anything else is known as a malpresentation, and unless you are experienced you should seek professional help.

A cow may give birth standing up or lying down. The first thing you will see is the water bag, then hopefully the two front hooves with the soles pointing down, then the nose. At the halfway stage the cow may rest and even walk about with half a calf showing. The cow will then expel the calf, which can fall to the ground if she is standing, and the umbilical cord will break. She should then turn around and start licking the calf, making soothing grunts to it. The calf can be on its feet within 30–60 minutes and will naturally find its way to the udder.

Next the placenta or afterbirth will be expelled; this should happen within 2–8 hours, but may take up to 24. The cow will normally eat the afterbirth; this is a natural animal reaction to hide any trace of the calf from predators.

When to intervene

If possible leave the cow to cope on her own. Nature knows best, but not always! If you see that the cow's waters have broken but there is no further progress, assistance maybe required. An internal inspection does not harm or distress the cow, and enables the way the calf is lying to be felt. As gently as possible, manipulate the calf into the correct position; experience or an experienced advisor are invaluable in such cases.

Dystocia is the word for an abnormal or difficult birth, and in most cases of dystocia the calf can be pulled by hand once its position has been corrected. When assisting a delivery, pull when the cow strains and rest when she does, and do not go on pulling continuously. There are various aids such as calving ropes that can be used, but a calf jack, which is a sort of winch, should only be used by a vet or very experienced calver.

You will probably need to intervene or call the vet if:

- ❏ The hooves protrude but the soles are facing up (probable breech birth).
- ❏ The hooves protrude but no further progress is seen for 45 minutes.
- ❏ Labour has lasted for more than three hours with no sign of a calf.
- ❏ Two people have been pulling for 20 minutes with no result.
- ❏ The calf has not got up in 2–3 hours and not suckled after six.
- ❏ The placenta does not clear by 24 hours after birth.

Once the calf has been born the mother starts to lick it, which encourages it to get up on its feet for its first suckle, of the colostrum. This is a form of milk produced by mammals in late pregnancy that contains essential antibodies and will keep the calf free from disease. The first dropping produced by the calf is known as meconium and will be soft, dark and rather smelly. In most cases all will be well and your beautiful newborn calf will be up on its feet and suckling in no time at all.

Castration

Unless you consider that they are going to grow up into champion bulls, male calves have to be castrated. There are three methods approved by Defra under the Protection of Animals Act:

- ❑ A rubber ring or other device, which can only be used in the first seven days of life.
- ❑ Bloodless castration, by a trained and competent stock-keeper, carried out by crushing the spermatic cords of calves less than two months old with a burdizzo.
- ❑ Castration by a verterinary surgeon using an anaesthetic.

It goes without saying that you should never attempt this procedure without first being given a demonstration.

Disbudding and dehorning

If you choose a horned breed, where possible allow your cattle to have natural horns or select a polled breed, but in order to minimize the risk of animals causing injury to each other you may decide to disbud your stock. The law is slightly vague in this area, but basically disbudding may be carried out in the first seven days of life by chemical cauterization, or with a heated disbudding iron up to about two months, without anaesthetic. Dehorning is a more painful procedure and an effective anaesthetic is essential.

Organic farming

There are strict regulations known as 'standards' that govern what organic farmers can and cannot do. Before you can call you cattle organic you are required by law to acquire a certificate from an Independent Certification body. In the UK the Soil Association can provide this.

Among the regulations, pesticides are severely restricted, chemical fertilizers are banned, animal cruelty is prohibited and a truly free-range life is promoted. The routine use of drugs, antibiotics and wormers is disallowed. Instead the farmer must use preventative methods such as moving animals to fresh pasture and keeping a smaller herd size. Genetically modified (GM) ingredients in food are also banned under organic standards. Unless the cow you buy is organic you will have to wait for you first calves before you can call you animals organic – a cow that is born 'unorganic' cannot become organic.

An inspector will come and inspect your farm or smallholding and talk through the necessary regulations if you want to go down this route. Further information can be found on the Defra and Soil Association websites.

Common ailments

As a rule cattle are healthy creatures and only suffer from minor problems, but it is as well to know what the most common ones are. Some ailments are notifiable diseases by law, which means that it is mandatory to inform your local animal-health department about them, and any on-farm deaths of animals over 30 months of age must be monitored by Defra as part of bovine spongiform encephalopathy (BSE) control measures. There are specific books dealing with cattle ailments.

Spotting that something is wrong as soon as possible is the key to keeping your herd healthy, and by simply glancing at every single animal once a day you will soon recognize if any are behaving oddly. Finding out locally if there are any problems, such as liver fluke, in the area will give you a head start.

Poisonous plants

Cattle generally avoid eating poisonous plants unless they are extremely hungry, but it is as well to know what they are:

- Yew
- Laburnum
- Rhododendron
- Ragwort
- Acorns if taken in very large quantities
- Bracken while growing; once it is fully out it seems to be tolerated, although the poison can be accumulative.

Parasites and skin problems

Worms are best avoided by constantly moving your cattle to clean pasture every three weeks as doing so will break the lifecycle of the parasites. If this is impossible or the animals are being housed inside over winter, you will need to administer anthelminthic wormers. There are different types of wormer and it is a good idea to vary them as some worms become resistant.

Ringworm is a fungal disease that affects humans as well as other animals and can be carried by them. The fungus can survive apparently forever on old fence posts and in buildings, and is almost impossible to eradicate. Round, bare, itchy patches appear on the skin. Ringworm can be treated with various proprietary preparations available from a local agricultural merchant.

Mange produces itchy patches on the skin where mites have burrowed in. A vet may need to look under a microscope to confirm the problem and provide advice.

Lice also produce itchy patches, and an affected animal's coat will look scruffy and unkept. There are plenty of treatments available to cure this problem. Housed animals and those in poor condition are most likely to be susceptible to lice.

Flukes are parasitic flatworms. There are a variety of anthelminthics to treat them, and a vet can advise you which is one suitable. Dose your animals in January and on land where fluke infestations are likely to be high; another dose in September/October is of benefit.

There are combined fluke and worming products on the market, but pay attention to the time of year they are administered.

Eye problems

New Forest Eye is a very contagious infection of the eye and if spotted early an animal should be isolated if possible. The symptom is a swollen, weeping eye that may be partially closed. There are a number of treatments available.

Notifiable diseases

Bluetongue is a disease that affects all ruminants, including sheep, cattle and deer. It does not affect horses or pigs. The disease is caused by a virus that is spread by certain types of biting midge. It is characterized by changes to the mucous linings of the mouth and nose, and the coronary band of the foot. Various restrictions are put in place if an outbreak is suspected. See Defra website for up-to-date information.

Bovine spongiform encephalopathy (BSE), or mad cow disease, is a fatal nervous disease of cattle that causes degenerative changes in the brain and other nervous tissues.

Bovine tuberculosis is an infectious disease caused by the bacterium *Mycobacterium bovis*, which can also infect many other mammals and cause TB in them. In the UK routine testing is carried out at the government's expense using the tuberculin skin test. The UK is divided into two testing areas. The South-west and East Sussex, where the majority of cases are currently found, are tested annually, and the rest of the country is on a four-yearly test programme. Cattle that react to the skin test are removed for slaughter and the cattle owner is compensated. The herd is placed under a movement restriction until all the remaining cattle have been tested clear.

Foot and mouth disease is an infectious disease affecting cloven-hoofed animals. It spreads swiftly and causes fever followed by the development of blisters chiefly in the mouth and feet. It is caused by a virus of which there are seven types, each producing the same symptoms and distinguishable only in a laboratory. In the UK the accepted policy is to stamp it out by slaughtering all affected stock and any others that have been exposed to the risk of infection. The owners of slaughtered cattle are compensated by the government.

The following are also notifiable diseases:

- Anthrax
- Bovine TB
- Brucellosis
- Enzootic bovine leucosis
- Warble fly

Schmallenberg virus is not a notifiable disease, but farmers are asked to contact their veterinary surgeon if they encounter cases of ruminant neonates or foetuses that are stillborn, or show malformations or signs of nervous disease.

Calf problems

Scour is the bovine word for diarrhoea. The best way to avoid it is to make sure that a calf has a good colostrum intake when newly born.

Navel-ill and joint-ill is a disease of young calves generally less than one week of age. It occurs as a result of infection entering the umbilical cord at or after birth. Signs of it are the calf having a high temperature and a swollen and painful navel. To avoid this problem, dip the umbilical cord in a disinfectant such as iodine as soon as the calf is born, and let it dry and drop off naturally.

Calves may also suffer from the following:

- ❑ Umbilical hernia
- ❑ Diphtheria
- ❑ Coccidiosis
- ❑ Salmonellosis
- ❑ Pneumonia

Adult problems

Mastitis is the inflammation of the mammary gland and udder tissue. It usually occurs as an immune response to bacterial invasions of the teat canal by sources present on a farm, in particular milking machines. The disease is treated with antibiotics.

Lameness may be caused by a foreign object lodged in a hoof, which may need trimming. Alternatively, it may be the result of footrot, in which case antibiotics are needed to effect a cure.

Hypomagnesaemia, or staggers, occurs in mature lactating dairy cows, usually within a few weeks of being moved onto lush grass in the spring. The onset is associated with a sudden fall in the plasma magnesium and calcium concentrations, and affected animals may be found dead or in severe trouble. Avoid the problem by supplementing magnesium in the feed.

Among other problems are leptospirosis, vibriosis, infectious bovine rhinotracheitis (IBR) and bovine viral diarrhoea (BVD) Type 1 and 2. Symptoms of these cattle diseases are embryonic death and poor conception rate.

Clostridial diseases: clostridial spores are widespread in the environment, particularly in soil and organic matter. Among the most common clostridial diseases are blackleg, botulism, red water, enterotoxemia and tetanus. Sudden death is often the first and only sign of these cattle diseases.

Bovine respiratory diseases (BRD), are the costliest of all cattle diseases, resulting in poor gains and a weakened immune system. Coughing, nasal discharge, fever and difficulty in breathing are among the symptoms of these cattle diseases.

Traumatic reticuloperitonitis, or wire-in-heart, is caused by ingesting foreign objects such as pieces of metal, nails and – more commonly in this day and age – the metal parts of Chinese lanterns that float away from parties and festivals. The sharp object penetrates the reticulum wall and infection spreads to the surrounding abdomen; in some cases a wire will penetrate the chest and cause infection of the outside of the heart. Treatment may involve surgery. Preventive measure include administering bar magnets to animals at one year old. These remain in the reticulum and hold any ferromagnetic objects on their surfaces.

There are many other ailments and diseases that the average farmer will hopefully never come across or have to deal with.

BREED PROFILES

ABERDEEN ANGUS

A tough producer of tender beef

Black hornless cattle have existed in north-east Scotland since the 18th century and were known colloquially as doddies and hummlies. The originator of the true Aberdeen Angus is thought to have been Hugh Watson of Angus in the early 1800s. His favourite bull, Old Jock, and best cow, Old Granny, supposedly gave birth to 29 calves, and the pedigrees of most Angus cattle alive today can be traced to them. Old Jock was given the number one in the Scottish Herd Book when it was founded. Another important breeder was Sir George Macpherson-Grant of Ballindalloch, on the River Spey, who spent his life refining the breed. The original animals were rather short in the leg and became more so until the mid-20th century, when breeders began to select for height.

This is a tough breed that is well suited to harsh northern winters and has a reputation for producing good-natured and straightforward animals. The cows' stayability, or ability to go on bearing calves, is legendary, and it is not unusual for cows to continue calving until up to 14 years of age.

Aberdeen Angus are one of the foremost beef animals in the world, producing well-marbled meat. They are used widely in cross-breeding as they mature early, calve easily and pass on the polled gene.

Aberdeen Angus calf

Aberdeen Angus cow

ORIGIN	TYPE	SIZE	HORNED OR POLLED
UK	Beef	Medium to large	Polled

COLOUR AND APPEARANCE

Predominantly black, but a dark red version does occasionally occur. Stocky appearance.

ANKOLE-WATUSI

The horns have it

You can see the ancestors of the Ankole breed carved onto the pyramids in Egypt, and East African tribes have always prized these animals with their immense horns. They were kept for their milk and as a sign of their owners' wealth, but were very seldom used for meat. Typically, a cow was grazed all day then brought home to her calf, which was allowed a brief suckle before the cow was milked by her owner. Mother and calf were then separated until morning, when the calf was allowed another suckle.

Ankole cattle were originally imported to European zoos during the late 19th and early 20th centuries for their curiosity value. Some were exported to American zoos, and when a few surplus animals became available to the public a few enterprising farmers started to produce herds. The Ankole Watusi International Registry was formed in 1983.

Ankole-Watusi steer

I could dance with you till the cows come home. Better still, I'll dance with the cows and you come home.

Groucho Marx

Beautiful lyre-shaped horns as long as 150 cm (60 in) are typical in the breed. It may be any colour, and either plain or spotted. The animals retain their own herd instincts. During the day the calves stay together with an 'aunt' cow looking after them, and at night the herd sleeps with the calves in the middle. Ankole milk is about 10 per cent butterfat and farmers utilize this by cross-breeding to boost fat levels. The breed tolerates huge ranges of temperature, with the horns acting either as radiators or for cooling as blood circulates through them.

ORIGIN	TYPE	SIZE	HORNED OR POLLED
Africa	Beef	Medium to large	Enormous horns

COLOUR AND APPEARANCE

Many varied colours and patterns. Rather angular and overshadowed by the horns.

AUBRAC

A mountain dweller from central France

Developed by Benedictine monks at Aubrac Abbey high up in the Massif Central, cattle of this breed spent the summer up in the mountains and walked to their pasturing ground on 25 May. They returned on 13 October to spend the winter under cover, and this tradition still continues today. At the *fête de la transhumance*, which takes place at the end of May, the cows are decorated with flags, bells and flowers, and paraded through the town of Aubrac. Up in the mountains the animals were formerly tended by shepherds who lived in basic huts called *burons*. They milked the cows out on the pasture and made Laguiole cheese that was matured in the *buronniers'* cellars. To while away the time they also made Laguiole knives – a kind of fold-up pocket knife or penknife – from the horns of the cattle.

Aubrac cow and calf

Aubrac bull

The Aubrac is a hardy, well-muscled beast that can withstand high altitudes and harsh climatic conditions, and produces well-marbled and flavourful beef on simple grass and hay. The breed is prized for its ease of calving, and Aubrac bulls are used on dairy and suckler herds to pass on this feat. This is a long-lived breed that is adaptable and known for its docile temperament.

ORIGIN	TYPE	SIZE	HORNED OR POLLED
France	Beef	Medium to large	Medium horns

COLOUR AND APPEARANCE
Fawn but can vary in shade to grey. The coat is darker on the shoulders and croup, particularly in bulls, in which it can run nearly to black. Well muscled.

AYRSHIRE

A worldwide dairy breed with elegant horns

Found all over the world, this is an old dairy breed that originated in the county of its name in south-west Scotland. Formed by crossing mainly shorthorns with several other breeds, it rivals Jerseys in milk production and yet is a hardy and long-lived animal.

Considered the ultimate economic dairy breed, Ayrshires are blessed with strong characters but mild temperaments and are efficient converters of forage. They are medium-sized and seem to adapt readily to all management systems, as well as being exceptionally healthy with good udder conformation and freedom from leg and hoof problems.

Ayrshire cow

Ayrshire horns were once a hallmark of the breed, reaching 30 cm (12 in) or so in length and being elegantly lyre-shaped – a spectacular sight when polished for the showring. Sadly, these days intensive dairy farming makes horns impractical and most cattle are now dehorned at birth.

ORIGIN	TYPE	SIZE	HORNED OR POLLED
UK	Dairy	Medium	Lyre-shaped, medium-sized horns

COLOUR AND APPEARANCE
Varying shades of red to mahogany-brown, with white patches with jagged edges all over the body.

BAZADAISE

A grey producer of first-class beef

Bazadaise are named after the town of Bazas just south of Bordeaux in France. A herd book has been in existence since 1895. Once used as working animals, this breed has now earned the coveted 'Label Rouge' for its low-fat, well-marbled and flavoursome beef. It is a tough breed suited to grazing in almost any climatic conditions, from those of the High Alps to the heat of lowlands.

During the late 20th century, breeding stock was exported to a good deal of Europe, including England, where the breed has gone from strength to strength. Particularly popular is its ease of calving, the calves being small at birth but developing fast. The cows are calm and good mothers, and most have a calving interval of less than 380 days. At birth the calves are a beige-wheaten colour, and they do not turn grey until they are three or four months old. Bulls are generally darker in colour than the cows, with a pale saddle, and all have the characteristic pale eyes and muzzle.

Bazadaise bulls are used for cross-breeding to improve conformation and weight gain.

Bazadaise cow and calf

ORIGIN	TYPE	SIZE	HORNED OR POLLED
France	Beef	Medium	Medium horns

COLOUR AND APPEARANCE
Ranges from dark to light grey, with pale patches around the eyes and muzzle. Calves are born wheaten and do not become grey until 3–4 months old.

BEEF SHORTHORN

Temperamentally calm and does well on poor grazing

According to the Shorthorn Society, this breed evolved over the centuries from Teeswater and Durham cattle from the north of England. In 1783 a farmer called Charles Colling found four cows called Duchess, Cherry, Strawberry and Old Favourite, and his brother Robert noticed that calves in a local market sired by a bull known as Hubback were superior to others, so he bought the bull for £8. These animals founded the Shorthorn breed.

Beef Shorthorn steers

This was originally primarily a dual-purpose breed, but gradually some herds tended more towards beef and others were bred for dairy. In 1958 Beef Shorthorns and Dairy Shorthorns were separated, and each breed now has its own section in the Herd Book. In order to improve muscling, Maine-Anjou blood was imported in 1976, but since 2001 outside blood has been banned.

Beef Shorthorn cows are good mothers, and as a result of their wide pelvic bone calve with ease. The calves reach puberty at an earlier age than those of other breeds. They do extremely well on mediocre grazing, as well as being temperamentally calm and easy to manage. The beef is well marbled, tender and has an excellent flavour.

ORIGIN	TYPE	SIZE	HORNED OR POLLED
UK	Beef	Medium	Horned – short

COLOUR AND APPEARANCE
Red and white, solid, spotted or brindled.

BELGIAN BLUE

Have these animals been to body-building classes?

Still the most popular beef animal in Belgium, this breed was created by crossing exported British Shorthorn and other bulls with native roan and pied cows. This was a dual-purpose breed until around the 1960s. Through selective genetic breeding towards size it became the extraordinary double-muscled creature seen today. Double muscling is a genetic mutation that represses the myostatin protein, resulting in greater muscle growth.

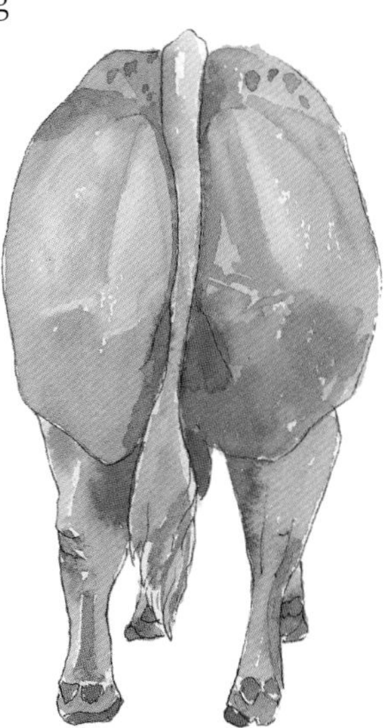

Belgian Blue cattle produce 20 per cent more meat per animal than most other breeds, but nothing is lost in the leaneness or tenderness of the meat. The only problem is that large beasts grow from large calves so that calving can be difficult, often necessitating Caesarian sections.

Known as the British Blue in the UK, this breed is prized for its quiet temperament and ease of handling.

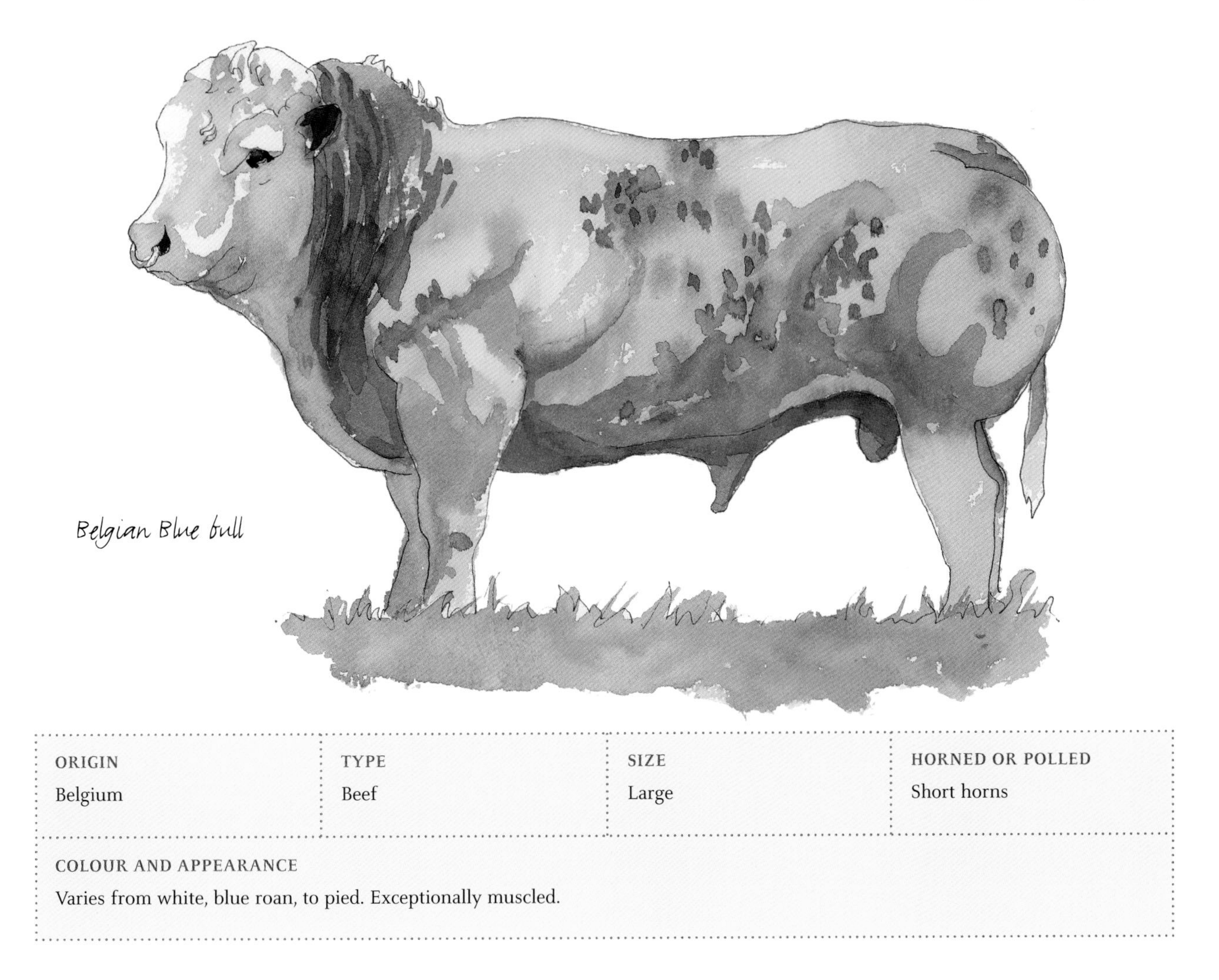

Belgian Blue bull

ORIGIN	TYPE	SIZE	HORNED OR POLLED
Belgium	Beef	Large	Short horns

COLOUR AND APPEARANCE

Varies from white, blue roan, to pied. Exceptionally muscled.

BELTED GALLOWAY

An ideal choice for rough ground

Although sharing many similarities with Galloways, Belted Galloways, or simply Belties, have been a separate breed for more than a century. The distinctive white belt around the body makes them easy to recognize, and they are currently enjoying a renaissance due to their suitability for rough grazing and use in conservation projects where the ground is unsuitable for machinery.

Belties are naturally polled, with a long, coarse outer coat that is shed in the summer and a soft undercoat that provides insulation in the winter, making them hardy enough to overwinter outside. Black is the most common and also most striking colouring, but dun and red Belties are also recognized by the Breed Society.

This is a calm and temperate Scottish breed, but the mothers are maternal and fiercely protect their calves from all threats, imagined or otherwise. As a result of being slowly grown, the beef is not ready until three years rather than 12–18 months, and is low in cholesterol and high in Omega 3. The meat can be slightly dark in colour and lean, but has even marbling that provides succulence and flavour.

ORIGIN	TYPE	SIZE	HORNED OR POLLED
UK	Beef	Medium to small	Polled

COLOUR AND APPEARANCE
Black, dun or red, with a central white belt right around the body. Rather woolly appearance.

BLONDE D'AQUITAINE

A breed for any climate

Blonde D'Aquitaine cow

Known mainly for its well-marbled, low-fat lean beef, the Blonde d'Aquitaine was once valued as a draught animal. Blonde cattle of one sort or another have been around in Europe since the 6th century, but were not officially recognized until 1962.

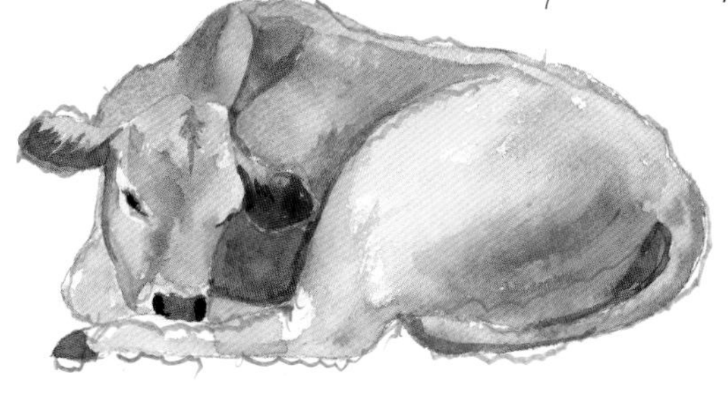

Blonde D'Aquitaine calf

Introduced to Britain in 1974 and known as the British Blonde, improvements to the breed were attempted by crossing it with Charolais, Shorthorn and Limousin cattle, but none proved to be any better than the original. The cattle are either naturally polled or horned, and have a reputation for being hardy animals that are tolerant of both hot and cold conditions.

This is a very docile breed known for its ease of calving and fast growth. Animals should have unbroken, wheaten-coloured, glossy coats, although paler and darker colours are acceptable. The expressive head should have a broad forehead with a triangular face, and the strong body should have an upright stance with well-developed joints and a deep chest.

ORIGIN	TYPE	SIZE	HORNED OR POLLED
France	Beef	Medium to large	Medium horns or polled

COLOUR AND APPEARANCE
Mainly golden-beige or wheaten, but also occurs in white and all shades to brown. In all cases the colour should be unbroken.

BRAHMAN

An ideal breed for very hot climates

The Brahman originates from *Bos indicus* cattle imported from India that are all characterized by having a large hump over the shoulders. Having been exposed over many centuries to inadequate feed, insects and parasites in India, through natural selection this breed has become well adapted to any hot climate and can survive where many other breeds would fail. These are the sacred cattle of the Hindu faith, and were the first beef breed developed in the US, where they played an important part in cross-breeding programmes.

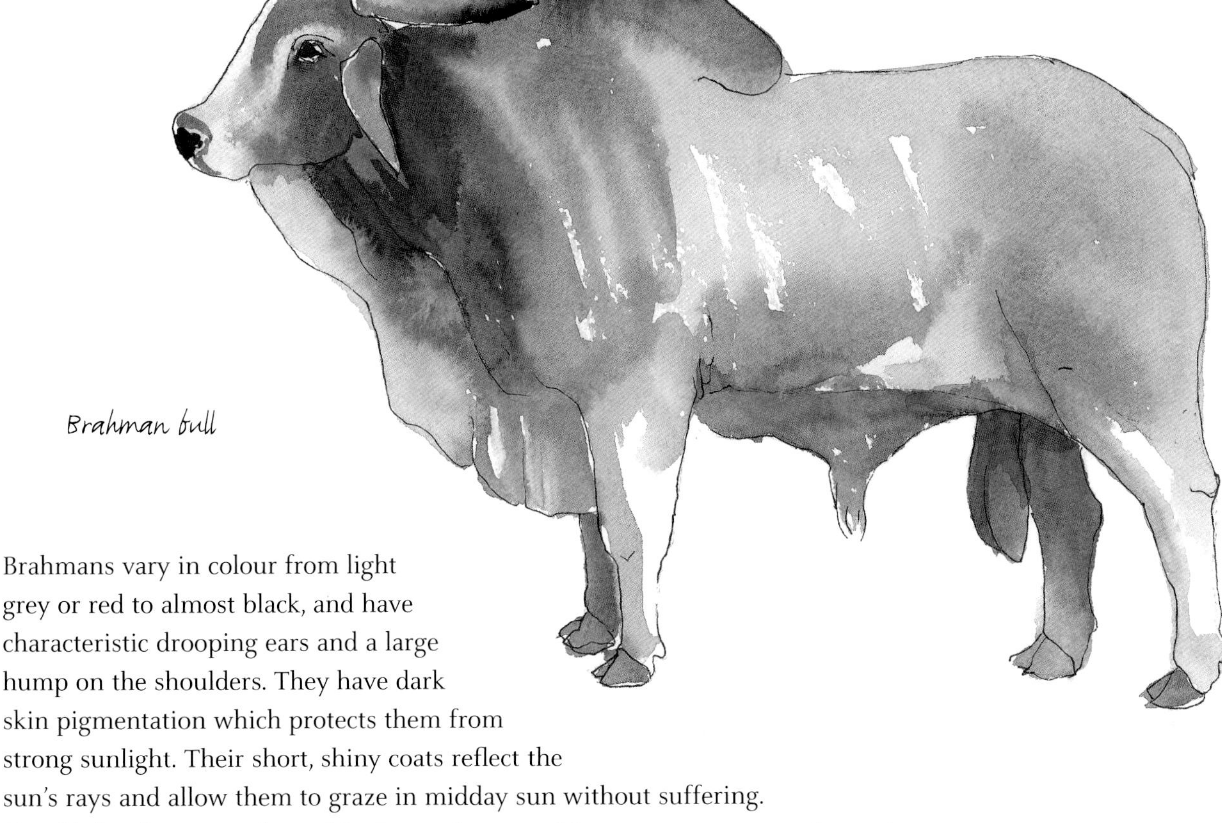

Brahman bull

Brahmans vary in colour from light grey or red to almost black, and have characteristic drooping ears and a large hump on the shoulders. They have dark skin pigmentation which protects them from strong sunlight. Their short, shiny coats reflect the sun's rays and allow them to graze in midday sun without suffering. Should they be exposed to cold winters they are able to grow a coarse outer coat.

Brahmans are able to utilize relatively low-quality feed as well as to travel further for it than other breeds, and this ability, along with ease of calving and a kind temperament, has made them a popular choice for cross-breeding.

ORIGIN	TYPE	SIZE	HORNED OR POLLED
India	Beef	Medium to large	Small to medium

COLOUR AND APPEARANCE
Many colours, from almost white to dark grey and black. Prominent hump and drooping ears.

BRITISH WHITE

An unusual choice for the smallholder

British Whites are the direct descendents of indigenous wild white cattle of Great Britain, and their history can be traced to Whalley Abbey in Lancashire during the 16th century. In 1918 the Park Cattle Society was formed, but it was not until 1946 that two forms of Park Cattle were recognized, the horned White Park (see page 144) and the polled British White.

British White bull

As a dual-purpose breed, the British White is impressively milky and is used as a suckler cow, its medium to large frame allowing it to calve easily. Due to its natural foraging skills it is often used in conservation and heathland-restoration projects, and it is naturally hardy and long-lived.

With their black or occasionally red ears and noses, these handsome beasts are ideally suited to the smallholder, being docile but hardy. They produce not only plenty of milk, but also a lean marbled carcass of excellent-quality beef.

ORIGIN	TYPE	SIZE	HORNED OR POLLED
UK	Beef and dairy	Medium to large	Polled

COLOUR AND APPEARANCE
White with black or, more rarely, red points.

BROWN SWISS

One quite at home in the mountains

Brown Swiss originated in the Alpine regions of Europe and are now found all over the world. They are one of the leading dairy breeds, particularly where cheese is the primary market. This is an ancient breed – it is claimed that its roots can be traced back to the Bronze and Iron Ages. Brown Swiss travelled with the Pilgrims to North America, from whence they spread worldwide. Sometimes known as the Bruhnvieh, the Brown Swiss was developed from this dual-purpose breed by selection of the best milkers.

Brown Swiss heifers

Brown Swiss are adaptable animals known for their tolerance of extreme temperatures, whether hot or cold, and able to graze at heights of 8,000 ft (2,440 m) due to their strength and durability. They seem particularly resistant to insects, and are renowned for their efficient conversion of forage.

The breed has been used for cross-breeding, passing on to its progeny its strong legs and feet, calm temperament and longevity, as well as its ease of calving.

ORIGIN	TYPE	SIZE	HORNED OR POLLED
Switzerland	Dairy	Medium to large	Medium horns or polled

COLOUR AND APPEARANCE
Light brown to chocolate-brown, with a creamy-white muzzle and dark nose.

CANADIENNE

Ideally suited to being a house cow

Cattle brought over by French immigrants from Normandy and Brittany during the 16th and 17th centuries formed the basis of the Canadienne breed. Although the breed was well suited to the Canadian climate, some farmers still thought there was room for improvement by crossing it with larger imported breeds, even though these were less adapted to local conditions. In 1895 a small group of breeders formed the Canadienne Cattle Breeders Association to protect the original bloodlines.

Canadienne cow

This is a small breed that is well suited to the smallholder as a house cow, being exceptionally calm and docile in character, as well as long-lived. The animals can adapt to a climate that is hot in summer but very cold in winter, producing a thick winter coat that is shed in summer. The cows produce good quantities of milk on average grazing. Calves are born pale coloured, but soon become the dark red to black of their parents, which generally have pale-coloured muzzles. There may occasionally be white on the udder, stomach and chest.

ORIGIN	TYPE	SIZE	HORNED OR POLLED
Canada	Dairy	Small	Polled

COLOUR AND APPEARANCE

Deep reddish-brown to black, with a pale muzzle.

CHAROLAIS

The world's top beef producer

Charolais bull

There are records of white cattle in the Charolles region of France as far back as AD 878. At that time they were used as multi-purpose beasts, not only producing beef and milk, but also being used as draught animals.

This large, muscular breed was one of the first continental breeds to be imported into the UK and USA. It revolutionized the beef industry with its superiority over native breeds. It adjusts well to varying climates and management systems, and is the leading breed for terminal sires of cross-bred beef cattle, as well as being used to improve existing breeds.

Charolais cow and calf

Bulls should be well muscled on good, strong legs to carry their weight, with a long, level back. Females should have a feminine appearance and not be too heavily muscled. Although the coat is generally creamy-white, the colour can vary to a light tan; broken-coloured cattle are not encouraged, although colour is of secondary importance to conformation. The coat is short in summer, but grows thick and curly for winter.

ORIGIN	TYPE	SIZE	HORNED OR POLLED
France	Beef	Large	Medium horns and, more rarely, polled in the US

COLOUR AND APPEARANCE
Creamy-white to wheaten, with a pink muzzle and heavily muscled body.

CHIANINA

The Italian giant of the bovine world

Pronounced 'kee-a-nee-na', this may be one of the oldest breeds of cattle in existence. The Roman agriculturist Columella and poet Virgil both wrote about it. Named after the Chiana Valley in Tuscany, until relatively recently it was used as a draught animal, being long in the leg, strong and extremely large – in fact, it may well be the largest breed of cattle in the world. Although as a draught animal it was bred for size and temperament, beef producers have maintained the size but improved the rate of growth.

Chianina cow and calf

The cows give birth to fawn-coloured calves, but the coat lightens with age until – like the coat of the adults – it is pale grey on black skin; the bulls are sometimes darker on their heads and shoulders. Due to the colour of the skin the breed is well adapted to hot climates, but can also tolerate colder ones. Chianina cows have unusually small udders and are not renowned for their milk production.

ORIGIN	TYPE	SIZE	HORNED OR POLLED
Italy	Beef	Large	Small horns

COLOUR AND APPEARANCE
White to pale grey, with black points and black skin. Very long legged.

CHILLINGHAM

The rarest of rare breeds

Considered one of the rarest animals on Earth numbering around 100, these are truly wild animals that are never handled. The single herd has been contained at Chillingham for over 700 years and has been genetically isolated for most of this time. In science inbreeding is known to lead to extinction but the Chillingham Wild Cattle continue to thrive.

The cattle breed throughout the year, the bulls adopting 'home territories' although allowing other cattle to graze it. As there is no specific rutting season the bulls are constantly competing with each other for the chance to mate and many of them show the battle scars.

Chillingham cow and calf

The cows still behave in the same way as wild cattle would have 1,000 years ago leaving the herd to find a quiet corner to give birth. The calf is left in its hiding place with the mother returning two or three times a day to feed it until, after a few days, the calf returns with its mother to join the herd. The other cows all show great interest and sometimes the mother has to protect her calf from being adopted by an 'auntie'.

ORIGIN	TYPE	SIZE	HORNED OR POLLED
UK	Beef	Medium	Large horns

COLOUR AND APPEARANCE
White with red ear-colour

CORRIENTE

A wiry beast with a good turn of speed

Named for their speed – *corriente* means 'running' in Spanish – these are descendents of the first cattle brought to South America, Florida and the West Indies by the Spanish as long ago as the late 15th century, and known as Criollo along with the Florida Cracker and Texas Longhorn. They were tough, small animals that were able to withstand the ocean crossing, as well as poor grazing when they arrived. In fact, they thrived and some became semi-feral, adapting well to the various regions and different climates they found themselves in. Occasionally they would be rounded up for beef, but this was not of good quality and by the 20th century demand for higher quality beef resulted in farmers fencing their land and introducing more profitable modern breeds.

Corriente heifer

Corriente cow

Corriente are popular for use in rodeos as they are fast, agile and tough. They calve easily without human help and are fiercely maternal; they are frugal eaters, and require even less water than some larger breeds. Their only down side seems to be that they are ace escapers, easily clearing fences or squeezing through small gaps in hedges.

The unique structure of the Corriente's muscles is more like that of wild game and similarly lean, but it can be just as tender as that of a beef animal with more fat.

ORIGIN	TYPE	SIZE	HORNED OR POLLED
Mexico	Beef	Small	Medium to large

COLOUR AND APPEARANCE
Very varied colouring.

DAIRY SHORTHORN

A versatile breed suited to all systems

According to the Shorthorn Society, this breed evolved over the centuries from Teeswater and Durham cattle from the north of England. In 1783 a farmer called Charles Colling found four cows called Duchess, Cherry, Strawberry and Old Favourite. His brother Robert noticed that calves in a local market sired by a bull known as Hubback were superior to others, so he bought the bull for £8. These animals founded the Shorthorn breed.

This was primarily a dual-purpose breed, but gradually some herds tended more towards beef and some were bred for dairy. In 1958 Beef Shorthorns and Dairy Shorthorns were separated, and each breed now has its own section in the Herd Book. Some breeders improved their herds by introducing outside blood from other breeds, but others preferred to keep their lines pure. This has resulted in some diversity of type.

This is a versatile breed suited to all types of production system. The cows are good forage converters able to produce quality milk in quantity in an economic manner. Like their cousin, the Beef Shorthorn (see page 36), they also have calm temperaments and are easy to manage.

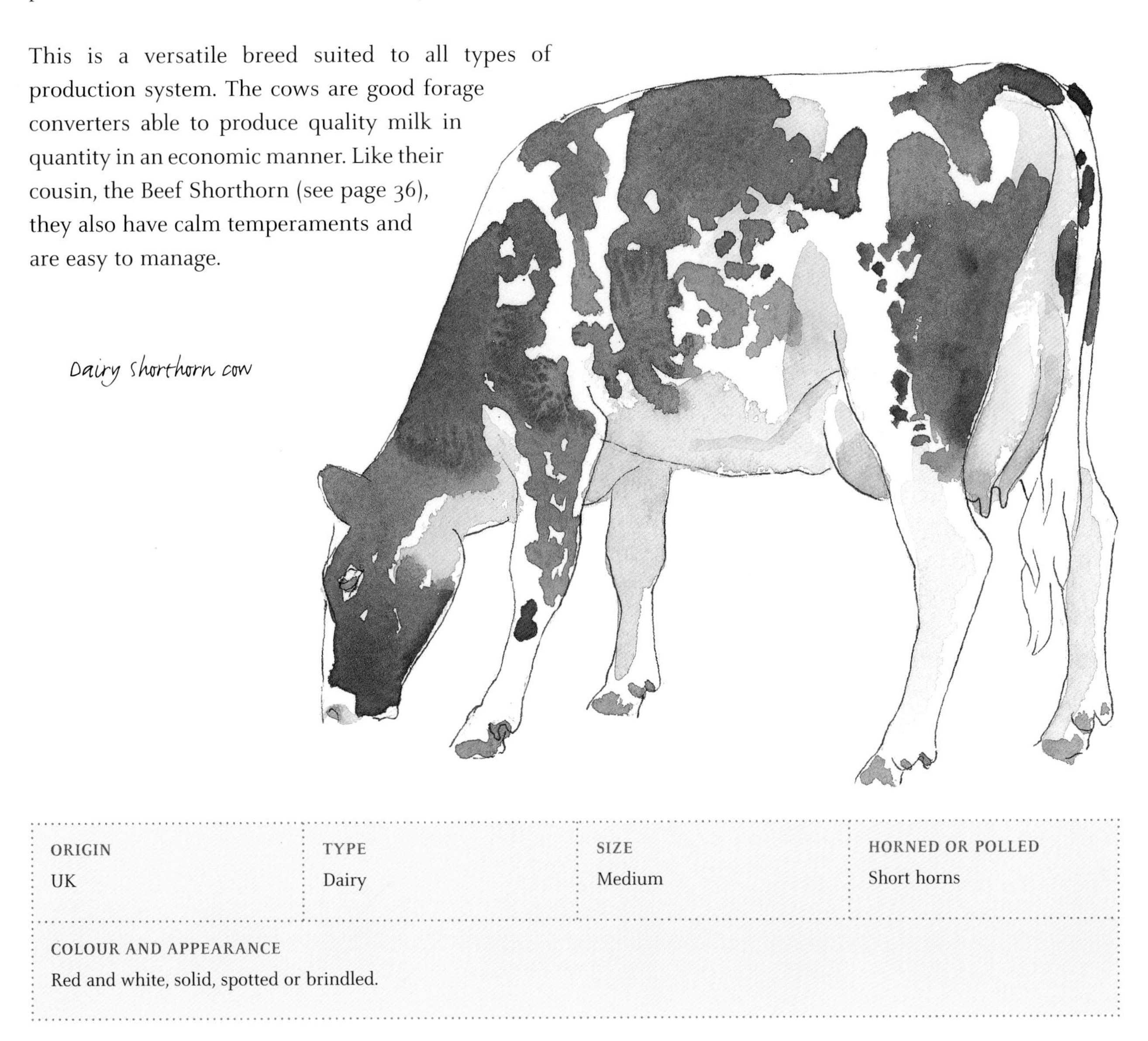

Dairy Shorthorn cow

ORIGIN	TYPE	SIZE	HORNED OR POLLED
UK	Dairy	Medium	Short horns

COLOUR AND APPEARANCE

Red and white, solid, spotted or brindled.

DEXTER

Small but perfectly formed

Mr Dexter of County Tipperary in Ireland created this unusual breed in 1750 by careful selection of the hardy mountain cattle in that area. They are thought to be directly descended from the mainly black cattle of the Celts. Although Kerry cattle (see page 90) also come from this source, they are not closely related.

Dexters are the smallest breed of cattle, measuring not more than 121 cm (4 ft) at the rump, and occur as two types: short legged and non-short legged. They are found in three colours, black being the most common, but also red and dun. Each colour should be solid over the entire body, although very small amounts of white hair are acceptable.

Dexter cow

The Dexter is a dual-purpose breed producing early-maturing beef of good quality and excellent flavour. The cows are extremely maternal and produce plenty of milk for their size, the fact that they may not produce as much as some purely dairy breeds making them popular with the smallholder. Even so, a good cow can raise its own calf as well as one bought in. If used purely as a dairy animal, a cow will produce an average of 10–12 litres (18–21 pints) of milk per day, and can continue to produce a calf a year for 14 years.

Gentle yet hardy, this truly is the smallholder's ideal animal.

ORIGIN	TYPE	SIZE	HORNED OR POLLED
Ireland	Beef and dairy	Small	Small to medium horns (though polled exist)

COLOUR AND APPEARANCE
Entirely black or entirely red or dun, with as little white as possible. Small bulls should be no taller than 121 cm (4 ft) at the rump, and cows 111 cm (3 ft 8 in).

DUTCH BELTED

Striking animals with a broad white belt

Seventeenth-century European noblemen, in particular Dutch ones, greatly admired the striking looks of the original belted cattle that came from Switzerland and Austria. They bred them specifically to encourage the band of white around the animal's middle. Their alternative name, Lakenvelder, means sheet or cloth, and refers to the white band. In some countries animals with this unusual marking are known as 'sheeted' cattle. The belt should extend from the shoulder to the hip and completely encircle the body.

Dutch Belted cow and calf

The breed is known for its grazing and foraging abilities, and produces plentiful milk on forage alone. Cows continue to breed well into their teenage years, producing a calf every year. In most cases the fairly light birth weight of the calves leads to normal deliveries without assistance.

According to the American Livestock Breeds Conservancy, by the 1970s the Dutch Belted breed was close to extinction. However, due to the efforts of a few dedicated farmers it survived in the US and its semen was exported back to the Netherlands. In the early 1990s there was a revival of interest in the breed, particularly among farmers interested in grass-based dairying, and it is now in a stronger position than it has been for some time.

Dutch Belted bull

ORIGIN	TYPE	SIZE	HORNED OR POLLED
The Netherlands	Dairy	Medium	Short horns

COLOUR AND APPEARANCE
Black with a broad white belt – occasionally red with a white belt.

FLORIDA CRACKER

Available in a colourway of your choice

Along with the Corriente and Texas Longhorn, the Florida Cracker is descended from Criollo cattle that in turn came from the original cattle imported into the Americas by the Spanish conquistadors. It is similar to the Texas Longhorn, but smaller and with shorter horns, and well suited to the scrubby lowland areas of Florida. These are tough, athletic-looking beasts that possess good resistance to the diseases and parasites that are found in tropical conditions.

Florida Crackers occur in nearly every colour pattern, including spots and brindle, the most common being black and reds in various shades, although very dark red is considered to be an indication of Shorthorn breeding and is not desirable.

In 1949 laws were brought in with regard to free-roaming livestock, requiring cattle in the US to be fenced. The combination of these laws and the introduction of larger beef breeds contributed to the decline of the Florida Cracker, which is now critically endangered. The formation of the Florida Cracker Cattle Association in the 1980s is helping to preserve the breed for posterity.

Florida Cracker cow and calf

ORIGIN	TYPE	SIZE	HORNED OR POLLED
USA	Beef	Small	Medium to large

COLOUR AND APPEARANCE
Varied colours, including spots and brindles. The most common is black.

FRIESIAN

The archetypal black-and-white dairy cow

During the 1800s, black-and-white cattle were imported from the Low Countries to the east coast of the UK, but importations were stopped in 1892 to try and prevent the spread of foot and mouth disease, which was endemic on the Continent. Friesians are closely linked to Holsteins and the various societies change their names on a regular basis from Holstein-Friesian to British Friesian – there seems to be some confusion between the two breeds.

Friesians are slightly smaller than Holsteins and carry more flesh. According to the British Friesian Breeders Club, they are renowned for their high fertility and strong conformation, which enable them to last for more lactations, thus spreading depreciation costs. They have retained some of their dual-purpose traits, with the potential to produce substantial quantities of milk as well as male calves that can be fattened to produce good-quality lean beef.

These are predominantly black-and-white pied animals and they are generally bred for this characteristic, although red pied animals are still found in the Netherlands.

ORIGIN	TYPE	SIZE	HORNED OR POLLED
The Netherlands	Dairy	Medium	Medium horns

COLOUR AND APPEARANCE
Black-and-white patches. Typical diary conformation.

GALLOWAY

A tough black breed with an ancient pedigree

Named for the area of south-west Scotland where it originated, the Galloway is one of the oldest breeds of beef cattle in the world. One of a few breeds of cattle that are naturally hornless, it has an exceptionally shaggy and thick coat with a dense undercoat, which makes it ideal for the cold, wet climate of its homeland. The outer coat is shed in the summer months, and in warmer climates the animals adapt and only grow their thick outer coat if needed.

Galloway cow and calf

The Galloway is similar in size to the Aberdeen Angus, but has lower calf weights at birth making for ease of calving. It is exceptionally docile and does well on poor-quality grazing, so is ideal for areas where the soil is unploughable, such as moorland and lowland heath. Due to its thick coat it can live happily outside during the harshest winters, and as it is not using energy to keep warm its winter feed needs are also lessened, making it economic to keep.

The well-marbled beef is juicy and flavoursome, and lacks the wasteful internal fat that surrounds the kidneys and heart in other breeds. This claim has been substantiated in United States Department of Agriculture (USDA) tests comparing the Galloway with eleven other breeds.

ORIGIN	TYPE	SIZE	HORNED OR POLLED
UK	Beef	Medium	Polled
COLOUR AND APPEARANCE Black, though the long coat may have a reddish tinge. Stocky build.			

GASCON

A tough breed from the Pyrenean foothills

Gascon cow

This tough breed originated in the Pyrenees and their foothills of Gascony, France, with their harsh climate and low-quality forage. As a result the animals are exceptionally hardy, very efficient at converting food and possess strong hooves developed over years of ranging in the mountains. The Gascon also has a strong resistance to heat, as well as black-rimmed eyelids that protect its eyes from bright mountain light. It grows a thick winter coat that – although short – has good water-repellant characteristics.

The cows calve easily without assistance, and the calves can gain an impressive 1 kg (2¼ lb) a day without supplementary feeding. They are a reddish colour at birth and do not get their grey coats until around four months of age. The bulls are also grey, but have black shading below the neck and stomach.

ORIGIN	TYPE	SIZE	HORNED OR POLLED
France	Beef	Medium	Medium horns

COLOUR AND APPEARANCE
Grey with black points. Calves are born reddish but become grey by four months.

GELBVIEH

An adaptable, dual-purpose breed

Literally meaning yellow cattle in German (and pronounced 'gelp-fee'), the Gelbvieh originated in Bavaria in the mid-18th century and was developed from several local breeds. It is one of the oldest German cattle breeds and was used as a draught animal as well as for its beef and dairy produce. In the 20th century red Danish cattle were introduced into the breed to improve milk production, but this is primarily a beef breed.

The breed was introduced to the US by imported semen from Germany and used in artificial insemination programmes. Females are registered as pure-bred at 7/8 Gelbvieh, and bulls must be 15/16. Animals should be a yellowish-red colour known as golden honey-red, but a strain of black Gelbvieh exists. Gelbviehs have fine coats that make them suitable for hot, dry climates, and they show a good resistance to ticks. Although the Gelbvieh was originally horned, polled cattle have been developed in the US.

This adaptable breed is reputed to possess superior fertility. The quiet-tempered cows make good mothers and calve easily, producing fast-growing calves.

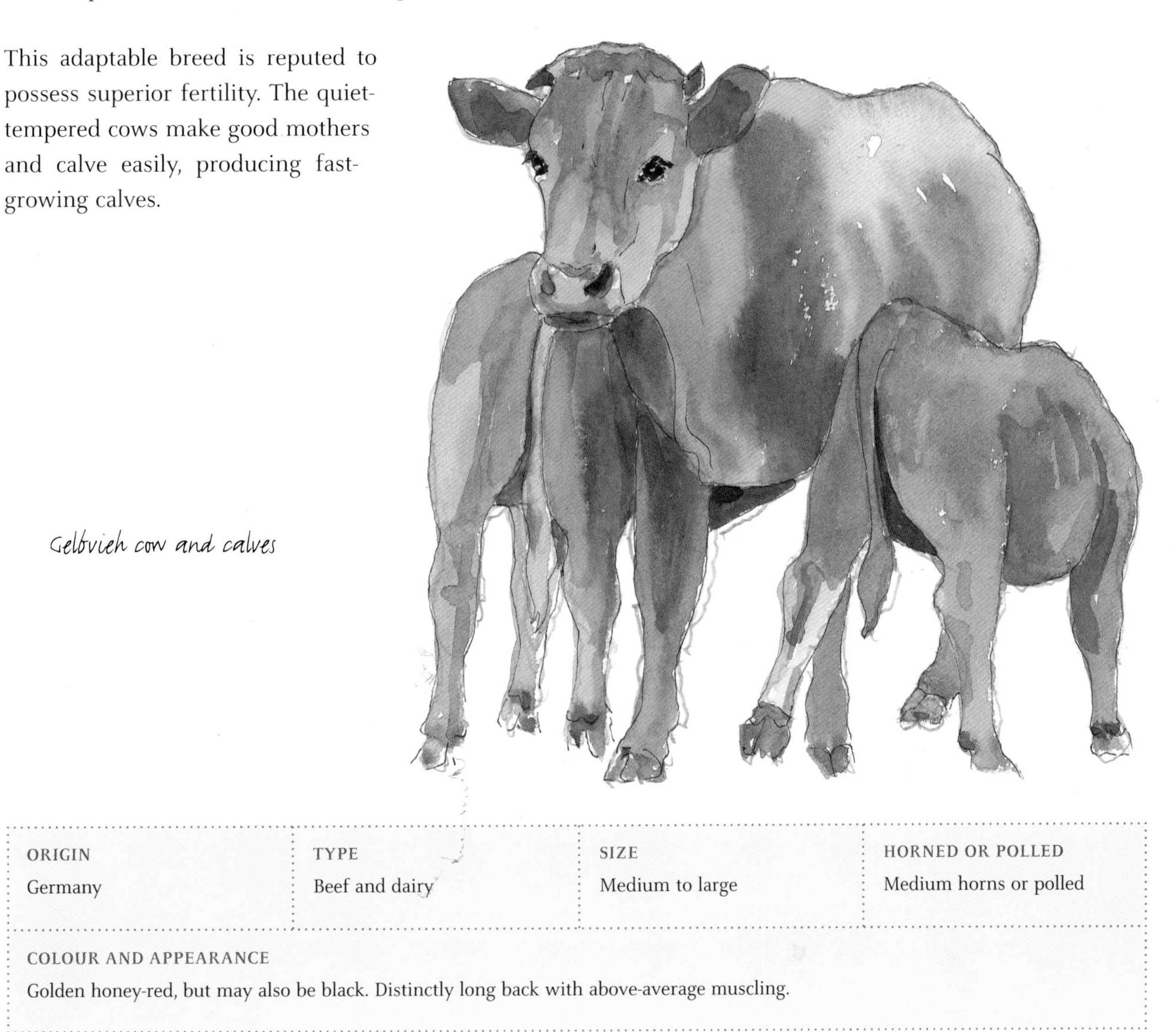

Gelbvieh cow and calves

ORIGIN	TYPE	SIZE	HORNED OR POLLED
Germany	Beef and dairy	Medium to large	Medium horns or polled

COLOUR AND APPEARANCE
Golden honey-red, but may also be black. Distinctly long back with above-average muscling.

GLOUCESTER

A milkmaid's saviour?

Best known perhaps for producing single and double Gloucester cheeses, this very handsome, dual-purpose breed is ideal for conservation and parkland management. It is a slow-growing breed that produces excellent, flavourful beef which now has a certification mark, signifying that only pedigree cattle can be sold as Gloucester beef.

Gloucester heifer

This is an ancient breed that was common in Gloucestershire and the surrounding areas as early as the 13th century, when it also supplied draught animals. Numbers gradually fell as other breeds became more popular, until in 1973 the Gloucester Cattle Society was formed to halt its decline. This the society has done very successfully, and Gloucester numbers – although this is still classed as a rare breed – are recovering and the breed is now recognized for its superb beef and cheese.

One moment of fame for the breed occurred in 1796, when a Gloucester cow called Blossom provided the first anti-smallpox serum after Sir Edward Jenner noticed that milkmaids appeared to be immune to this serious disease.

A cow is a very good animal in the field, but we turn her out of a garden.

Samuel Johnson

Gloucester cow

ORIGIN	TYPE	SIZE	HORNED OR POLLED
UK	Beef and dairy	Medium	Medium horns

COLOUR AND APPEARANCE
Black-brown body with black head and legs. White finching stripe along the spine, and white tail and belly. Slightly roman nose and fine horns that turn up at the tips, which are black.

GUERNSEY

A golden producer of the richest milk

These pretty dairy cattle have been classed as a breed of their own since 1700. They are thought to have developed from Isigny cattle from Normandy and Froment de Leon from Brittany. By 1789 importing foreign cattle onto Guernsey was banned to maintain the purity of the breed.

Guernseys are renowned for the quality of their golden-coloured milk, which is very high in butterfat and protein, and the possible health benefits from it, which include protection from Type 1 diabetes and heart disease. The milk contains more calcium than that of any other breed. It is rich in vitamin A and beta-carotene, which gives it its characteristic colour.

Guernsey cow

The export of cattle and semen are economically important for the island of Guernsey and the breed is now found in many parts of the world, although it is declining due to the popularity of the Holstein.

Guernseys are larger than Jerseys but equally docile, although the bulls have a reputation for being exceptionally aggressive. They are pretty cows with their pale orange – always called 'golden' – coats and skewbald markings.

ORIGIN	TYPE	SIZE	HORNED OR POLLED
Channel Islands	Dairy	Large	Medium horns
COLOUR AND APPEARANCE			
Golden-brown with white patches.			

HEREFORD

One of the oldest breeds of beef cattle

Native to the UK, the Hereford is one of the oldest beef breeds. The characteristic white face is a dominant trait that bulls pass on to all their progeny, whatever breed they may be crossed with. Originating in the Welsh Marches, the breed is thought to have developed from the small red cattle of Roman Britain and crossed with larger Welsh breeds. Herefords were imported into the US in 1817, and a breeding programme to create a hornless Hereford was begun.

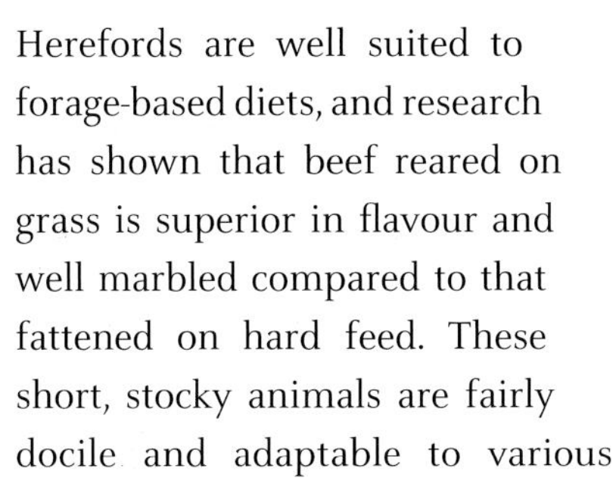

Hereford cow

Herefords are well suited to forage-based diets, and research has shown that beef reared on grass is superior in flavour and well marbled compared to that fattened on hard feed. These short, stocky animals are fairly docile and adaptable to various production systems and climatic conditions. Another point in their favour is ease of calving, and Herefords have traditionally been used by dairy farmers to capitalize on this trait.

All Herefords should have a white head and crest, white dewlap and throat, and white underside. The end of the tail and lower legs should also be white, contrasting with the deep rich red of the rest of the coat.

ORIGIN	TYPE	SIZE	HORNED OR POLLED
UK	Beef	Medium	Medium horns or polled

COLOUR AND APPEARANCE
Rich red with a white face and underbelly. Stocky, with rather short legs.

HIGHLAND

One for the open moor

Highland heifer

It is the double coat that makes these animals so suitable for the rugged Highlands of Scotland. The soft, downy undercoat provides warmth, and the long, straggly overcoat is slightly oily and therefore waterproof, allowing the animals to winter outside. In hot weather the outer coat can be shed. Highlands have magnificent horns; as a rule the cows' horns grow outwards and curve up, whereas the bulls' horns curve out and forwards. It takes several years for the horns to reach their final length.

Herds of Highlands are known as 'folds', as in ancient times the herder would put his animals into a fold at night to protect them from wild animals and the climate. It was this breed that was driven all over the country by the Scottish drovers. They lived in the Highlands and in the spring would visit the many small farms to bargain for cattle until they had formed a fold of at least 100 beasts. They would drive these using ancient drove roads to sell in the more prosperous south.

Highlands occur in a variety of colours, ranging from black, brindled and red, to yellow and dun. They are docile, gentle creatures – even the bulls are generally good natured, although like all bulls they should be respected. They are long-lived and generally produce more calves that other breeds, continuing to produce until 20 years of age.

ORIGIN	TYPE	SIZE	HORNED OR POLLED
UK	Beef	Medium	Large horns

COLOUR AND APPEARANCE
Mainly dun, but also black, red or brindled. Stocky build, with a long, curly coat.

HOLSTEIN

The most prolific milk producer of all

Originating in what was once the two provinces of north Holland and Friesland and is now the Netherlands, Holsteins developed from the black-and-white animals of migrant European tribes around 2,000 years ago.

Although similar looking to Friesians, Holsteins are larger and have a separate entry in the *Herd Book*. They are outstanding milk producers – a single cow can produce up to 8,600 litres (15,135 pints) a year, and keep this up over an average of three lactations. Although this amazing quantity outdoes most other breeds on a volume basis, the milk has a lower percentage of butterfat, which is needed in the making of cheese.

A famous Holstein cow called Pauline Wayne belonged to William Taft, 27th President of the US, and spent her life grazing on the White House lawn. She was the last presidential cow to live at the White House as this tradition became obsolete.

ORIGIN	TYPE	SIZE	HORNED OR POLLED
The Netherlands	Dairy	Large	Medium horns

COLOUR AND APPEARANCE
Black and white in patches, and occasionally red and white.

IRISH MOILED

Ideal dual-purpose animal for smallholders

This rare and ancient breed is thought to have arrived in Ireland with the Vikings in the 8th or 9th century, although Irish legends often refer to red, white-backed cattle, and apparently skeletal remains have been dated to AD 640. The name Moile is gaelic for 'little round' or 'mound', and refers to the dome on the cattle's head. The breed can range in colour from almost entirely white with red ears, to almost completely red. However, the Irish Moiled Cattle Society prefers animals to be predominantly a solid rich red with a broad white stripe running down the full length of the back (finching), and white underparts and tail.

The breed reached a peak in the early 19th century, but slowly declined due to the introduction of more popular dairy and beef breeds. It almost became extinct in 1970, but the Rare Breeds Survival Trust recognized it as endangered and added it to the critical list. A few enthusiasts became breeders, and this delightful and adaptable breed was saved.

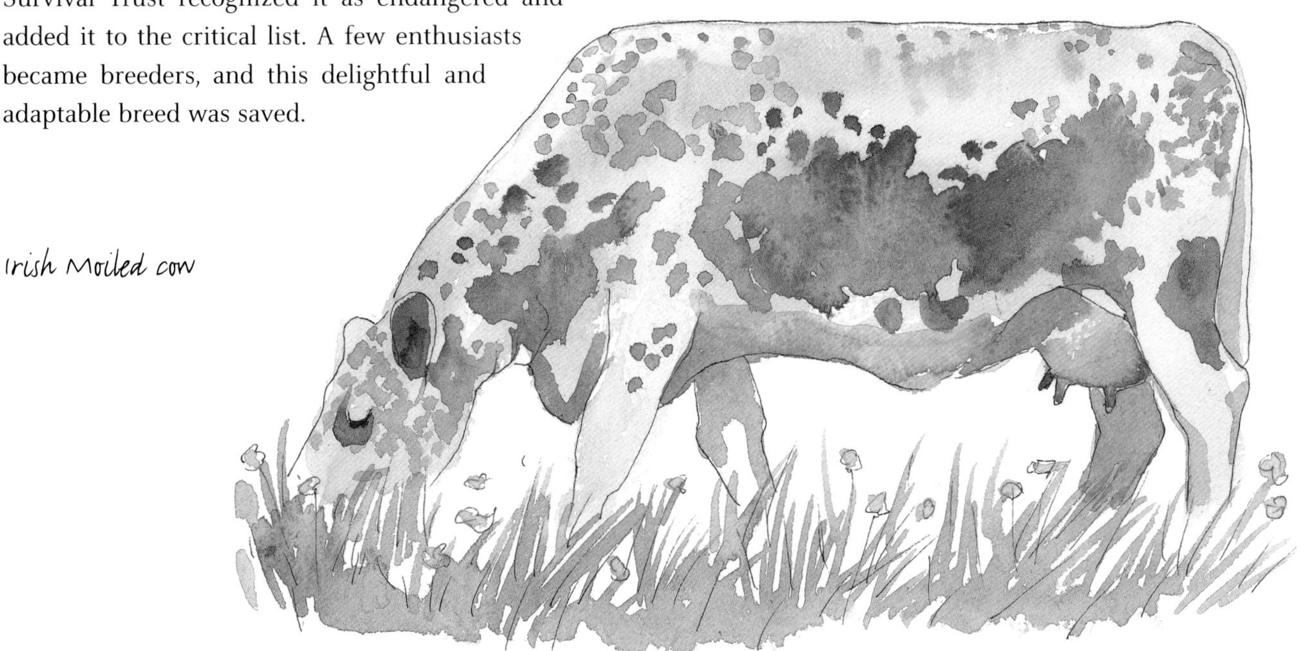

Irish Moiled cow

Economical to run and with a placid temperament, this really is an all-round ideal animal for the smallholder. The cows produce up to 5,000 litres (8,800 pints) of milk and a calf a year until at least ten years of age – and it is not unheard of up to 15. The beef is well marbled and tender. These cattle are also browsers and make the most of whatever they can get, so are useful in conservation grazing schemes. Moileys grow a good winter coat and can therefore winter outside.

ORIGIN	TYPE	SIZE	HORNED OR POLLED
Ireland	Beef and dairy	Medium	Polled

COLOUR AND APPEARANCE
Mainly red or roan, with a white finch back and underbelly.

JERSEY

The archetypal dairy cow

Jersey cattle have occurred on the island of Jersey, Channel Islands, for more than 1,000 years. The purity of the breed is maintained by an import ban on the island that has been in place for 150 years. Jerseys have been documented in the UK from around the 1750s, when they were known as Alderneys, and the breed reached the US by the 1850s. The Jersey Cattle Society was formed in 1878, and one of the oldest registered herds is that of HM The Queen at Windsor.

The Jersey is a smallish dairy cow that produces high-quality milk, and is the second most popular dairy breed after the Holstein. The milk contains 18 per cent more protein, 20 per cent more calcium and 25 per cent more butterfat than regular milk. A cow can average 5000–9000 kg of milk in one lactation or 15–20 litres per day, and being good calvers the animals can be crossed with larger breeds. The cows also mature earlier than those of most other breeds, and are able to calve from the age of 20 months rather than the average of 24.

Pure-bred Jerseys can range in colour from light fawn through mousy-grey to almost black (known as mulberry); they can have white patches although a plain colour is preferred. All Jerseys have a black nose, white muzzle and black switch. Their black hooves are hard and less susceptible to lameness than those of other breeds. The Jersey is versatile when it comes to temperature and can adjust to cold by growing a thick coat, but also thrives in heat.

Jersey bull

This pretty breed is known for its kind temperament, although the cows can be a little nervous and the bulls, like those of most breeds, are renowned for being aggressive. The Jersey is an ideal choice for the smallholder.

ORIGIN	TYPE	SIZE	HORNED OR POLLED
Channel Islands	Dairy	Small to medium	Medium horns

COLOUR AND APPEARANCE
Shades of light fawn and cream, to darker chestnut, to almost black, always with a black nose with a white border. Black switch. Dished face with exceptionally large, doe-like eyes. Hollows in front of the hip bones.

KERRY

An ideal choice for the smallholder

One of the oldest European breeds, the Kerry is thought to originate from the shorthorns brought to Ireland by Neolithic man. The breed is also claimed to be the first to be primarily bred as a milk producer.

Kerrys are hardy, long-lived and ideally suited to poor-quality grazing. In fact, the breed was once known as 'the poor man's cow', as nearly every small Irish farm would have kept one as a house cow.

Kerrys are jet-black with horns that also have black tips. Their coats grow thick in the winter and they happily winter outside. Being small three can be kept on the equivalent pasture to two of a larger breed, and due to their agility they do less damage to the ground than other breeds. The cows calve easily although the calves fatten slowly, forming excellent-quality beef.

Butterfat in Kerry milk is found in smaller drops than in the milk of other breeds, making the milk easier to digest and therefore ideal for feeding to toddlers and invalids, but it also makes excellent cheese.

Kerry cow

All is not butter that comes from the cow.
Proverb

ORIGIN	TYPE	SIZE	HORNED OR POLLED
Ireland	Dairy	Small to medium	Medium horns

COLOUR AND APPEARANCE
Black but occasionally red – some white markings are acceptable.

LIMOUSIN

Could this be the same breed as that painted by cavemen?

Originating to the west of the Massif Central, in France, this may well be one of the most ancient cattle breeds as it looks very similar to animals depicted in the cave drawings of Lascaux, near Montignac, which are estimated to be 20,000 years old. This is a wet area with a harsh climate and poor soil that over the centuries has made the breed unusually sturdy and healthy. Once used as a draught animal, it is now farmed exclusively for beef.

Limousin bull

Limousins did not arrive in the UK until 1971, but very soon proved their worth by taking the number one beef AI position from the Hereford in 1986; they still hold this honour today. Limousin bulls are exceptionally fertile and pass on their conformation to their progeny. The cows, being relatively light framed, calve with ease, and the fact that they are generally smaller than other continental suckler dams allows for increased stocking.

Limousin cow and calf

ORIGIN	TYPE	SIZE	HORNED OR POLLED
France	Beef	Medium	Medium horns or naturally polled

COLOUR AND APPEARANCE
Golden-red, lighter under the stomach and around the eyes and muzzle; occasionally black.

LINCOLN RED

Exceptionally adaptable to grazing and climatic conditions

The Vikings are credited with bringing the ancestors of this breed to British shores, and records of Lincolnshire cattle exist from the late 17th century. During the 18th century these were crossed with cherry-red Durham and York Shorthorns to produce Lincolnshire Red Shorthorns. In 1939 work was carried out to create a polled animal. Red-and-black Aberdeen Angus bulls were used, and eventually the first polled Lincoln Red bull was granted a licence by the Ministry of Agriculture. The word Shorthorn was dropped and in 1960 the breed became simply the Lincoln Red.

Lincoln Reds are exceptionally efficient food convertors, being adaptable to grazing and various climatic conditions. A leg at each corner gives them enviable conformation, and the fact that they are also hardy, docile and long-lived allows them to stand up to their continental competition.

Lincoln Red cow and calf

ORIGIN	TYPE	SIZE	HORNED OR POLLED
UK	Beef	Medium	Polled

COLOUR AND APPEARANCE

Rich mahogany-red.

LONGHORN

One that would grace any pasture

This is a very old, traditional breed from the north of England, not to be confused with the Texas Longhorn of the USA. At one time in the 18th century it was a very popular breed – the horns were valued for the making of buttons, cutlery handles and spoons. By the end of the 19th century materials other than horn were being used and the housewife demanded leaner meat, as a result of which the Longhorn nearly became extinct. Thankfully, the Rare Breeds Survival Trust saw the potential of the breed and saved it – it is now out of danger and has become popular with small breeders as a result of its elegant looks and all-round qualities.

Considered an excellent suckler cow, the Longhorn calves with ease and produces milk with a high butterfat content to feed her progeny. The milk is also used in the making of Stilton and other well-known hard cheeses. This is a long-lived, hardy breed known for its docility, which produces well-marbled beef that is both flavourful and succulent.

Longhorns' coats should all have a white line down the back and a white patch on the thigh. The most common and most popular colouring is red brindled – animals that are white with no red hair at all are discouraged by the Longhorn Society.

Longhorn cow

ORIGIN	TYPE	SIZE	HORNED OR POLLED
UK	Beef	Medium to large	Large horns

COLOUR AND APPEARANCE

Many shades, but red brindled is the most common. White line on the back.

LUING

A hardy island dweller

The rugged Isle of Luing off the west coast of Scotland is where this breed was developed by the Cadzoa brothers in 1947. They selected the best they could find of Beef Shorthorns and Highland Cattle. The first-cross heifers were then bred to a Shorthorn bull. Careful breeding continued until in 1965 the Luing was recognized as a breed in its own right.

This is a very hardy breed that can withstand the harshest of winters by growing a really thick winter coat which is shed in the summer. It is well suited to rough grazing and is ideal for places where grazing conservation schemes are in use. As a rule it is placid and easy to manage, but allowances must always be made for cows with calves. Cows are exceptionally long-lived, producing a calf a year for at least 10 and often many more years.

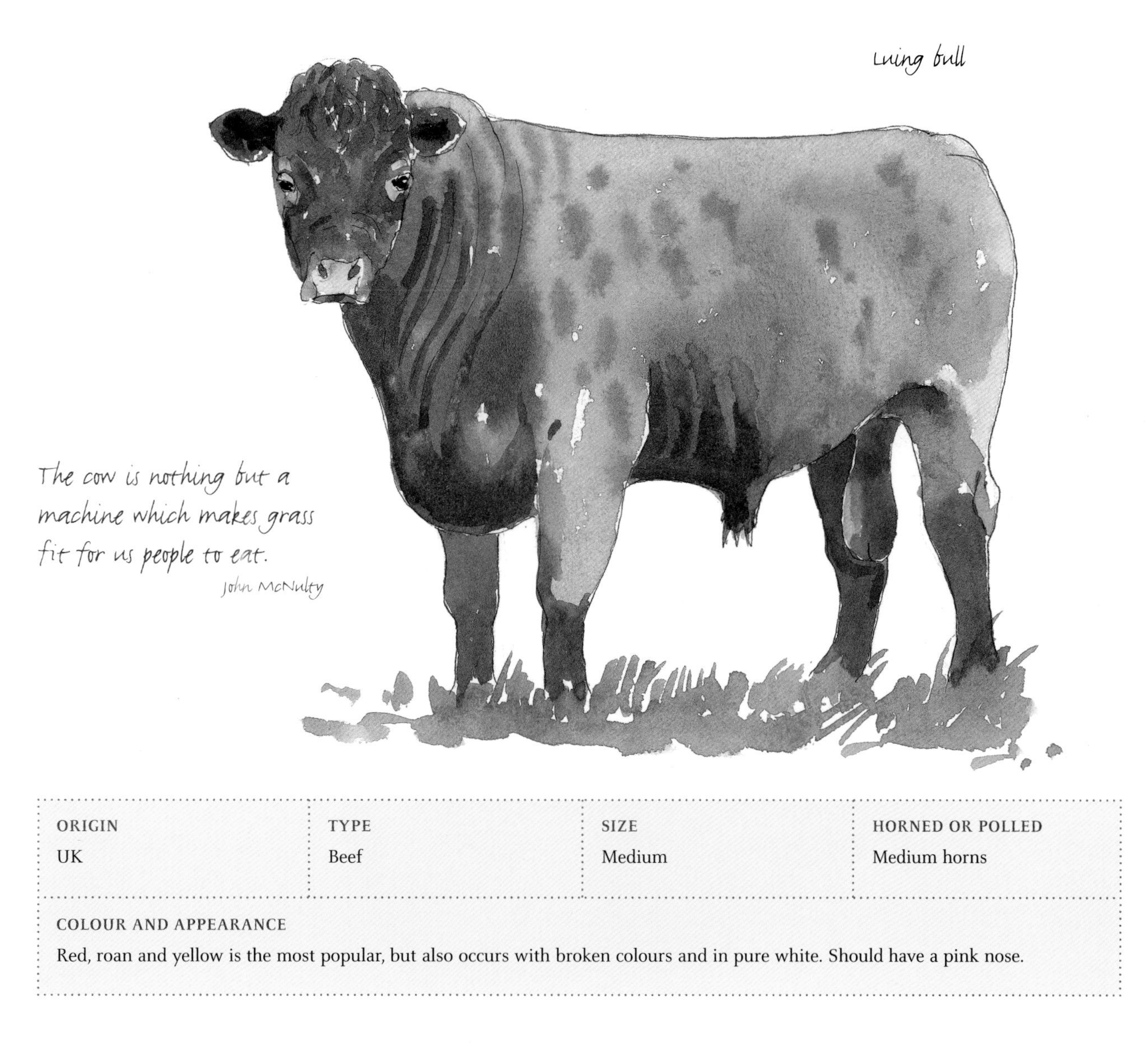

ORIGIN	TYPE	SIZE	HORNED OR POLLED
UK	Beef	Medium	Medium horns

COLOUR AND APPEARANCE

Red, roan and yellow is the most popular, but also occurs with broken colours and in pure white. Should have a pink nose.

MAINE-ANJOU

A gentle giant for the best of both worlds

Originating in the beef-producing north of France, this large breed is still sometimes used as a dual-purpose animal, though its primary use is for beef. It was developed from a crossing of the local Mancelle breed with Durham Shorthorn cattle imported from England, and their influence is still seen in the striking red-and-white colour of the animals' coats.

Maine-Anjou cow and calf

As a dual-purpose breed this was an ideal choice for the smallholder, who was able to use the cows for milk production and the bull calves for beef. In many herds half the cows were milked, and the other half raised two or three calves each annually – a practice that continues in Europe today. Although exceptionally large, these are also outstandingly docile creatures and very easy to handle despite their size. They are also very efficient food convertors, resulting in fast growth. They are long-lived, and have good, strong feet and thick hides, so that they can tolerate both cold and hot climates.

ORIGIN	TYPE	SIZE	HORNED OR POLLED
France	Beef	Large	Medium horns or polled

COLOUR AND APPEARANCE
Dark mahogany-red with white patches, particularly on the chest, hind legs and tail.

MARCHIGIANA

A large breed adapted to varying climates

No one quite knows the true origin of the Marchigiana, except that it originated in the Marche area of Italy to the north of Rome. This is a particularly rugged area with poor grazing and a harsh climate that varies from the heat of summer to cold, wet winters. Most likely Chianina blood was introduced at some point because this breed is very similar in looks, though slightly shorter in the leg.

The Marchigiana is large and well muscled, with a short but dense coat varying in colour from totally white to a light steely-grey. The skin is pigmented and the points, eyes, muzzle and switch are dark grey to black. The horns have black tips, and as in many breeds curve forwards in the bulls and upwards in the cows.

The cows possess strong maternal instincts and give birth to small calves that grow fast; in the best cases calves can gain up to 2 kg (4½ lb) a day. They also have the desirable reputation for being calm and friendly, making handling stress free. The Marchigiana breed constitutes approximately 45 per cent of all white cattle breeds in Italy.

Marchigiana heifers

ORIGIN	TYPE	SIZE	HORNED OR POLLED
Italy	Beef	Large	Small horns or polled

COLOUR AND APPEARANCE

White to light grey with dark points. Well muscled.

MEUSE RHINE ISSEL

A popular choice for cheese and ice-cream making

This breed originated in both the Netherlands and Germany. In the Netherlands it comes from the region of the three rivers that formed its name (in Dutch, Maas-Rijn-Ijssel) from the late 19th century. In Germany it is known as the Rotbunt (meaning red pied) and comes from the regions of Westfalia, Rhineland and Schleswig Holstein.

Originally this was a dual-purpose breed, but most herds are now entirely dairy. The milk contains a high kappa casein-B content, which is needed for cheese making and is also ideal for ice cream, although the beef is also excellent.

Meuse Rhine Issel are medium-sized, red-and-white animals that are solidly built and have a thick skin that enables them to withstand poor weather conditions. They are docile, adaptable and hardy, with good disease resistance, and do well on poor or rough pasture. The cows have a relatively short gestation period and calve easily, producing active calves with a strong will to live that take readily to bucket feeding.

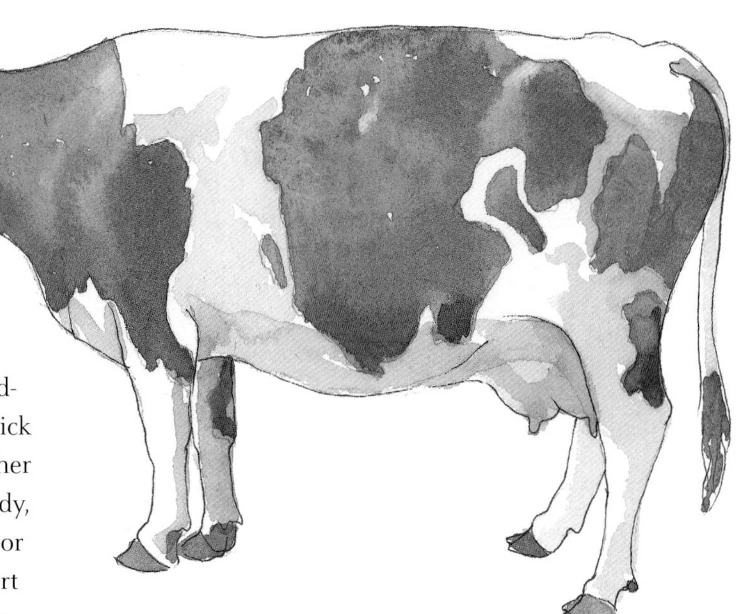

Meuse Rhine Issel cow

Meuse Rhine Issel cow and calf

ORIGIN	TYPE	SIZE	HORNED OR POLLED
The Netherlands and Germany	Dairy and beef	Medium	Medium horns

COLOUR AND APPEARANCE
Red-and-white pied. Solid looking.

MONTBELIARDE

A popular choice for mountain dwelling

Coming from the mountainous Jura region in the east of France, with its continental climate of hot summers and cold winters, this is not surprisingly a very hardy breed. It is also blessed with the strong legs and feet necessary for intensive dairy farming – it is, in fact, the second most populous dairy breed in France and is now found all over the world. The history of the Montbeliarde dates back to the beginning of the 18th century, when Swiss farmers from the Bernese Oberland moved to the principality of Montbeliard, bringing their livestock with them. The breed was officially recognized with its own herd book in 1889.

Montbeliarde milk is renowned for its high protein and kappa-casein B content, which is needed in cheese making. No less than nine cheeses quote Montbeliarde milk in their list of requirements, Comte, Mont d'Or and Reblochon being just three. This is truly a dual-purpose breed: the calves grow rapidly and the cull cows also produce good carcasses with no excess fat.

Montbeliarde cow

Sacred cows make the best hamburgers.
Mark Twain

ORIGIN	TYPE	SIZE	HORNED OR POLLED
France	Dairy and beef	Medium	Medium horns

COLOUR AND APPEARANCE
Red-and-white pied.

MURRAY GREY

Top-class beef from the antipodes

This breed appeared by accident in Australia's Upper Murray Valley in New South Wales. In around 1905 a breeder noticed that when he put a roan Shorthorn cow to his Aberdeen Angus bull, the calves were always grey and not roan as expected. Soon it was realized that these grey cattle were unusually quick to grow and were superior converters of feed. Their popularity grew and in 1960 the Australian Murray Grey Society was founded.

Murray Greys are very adaptable creatures that take happily to most climates, their grey skin protecting them from sunburn, but they are equally at home in colder, harsher conditions. Among their other attributes is ease of calving – calves are quick to be on their feet and suckling. They have a gentle disposition and are naturally polled. Their beef is world renowned for its excellent marbling and high yield, with less back fat than that of many other breeds. They are also highly efficient food convertors, with the steers finishing in up to half the time with half the food input of other breeds.

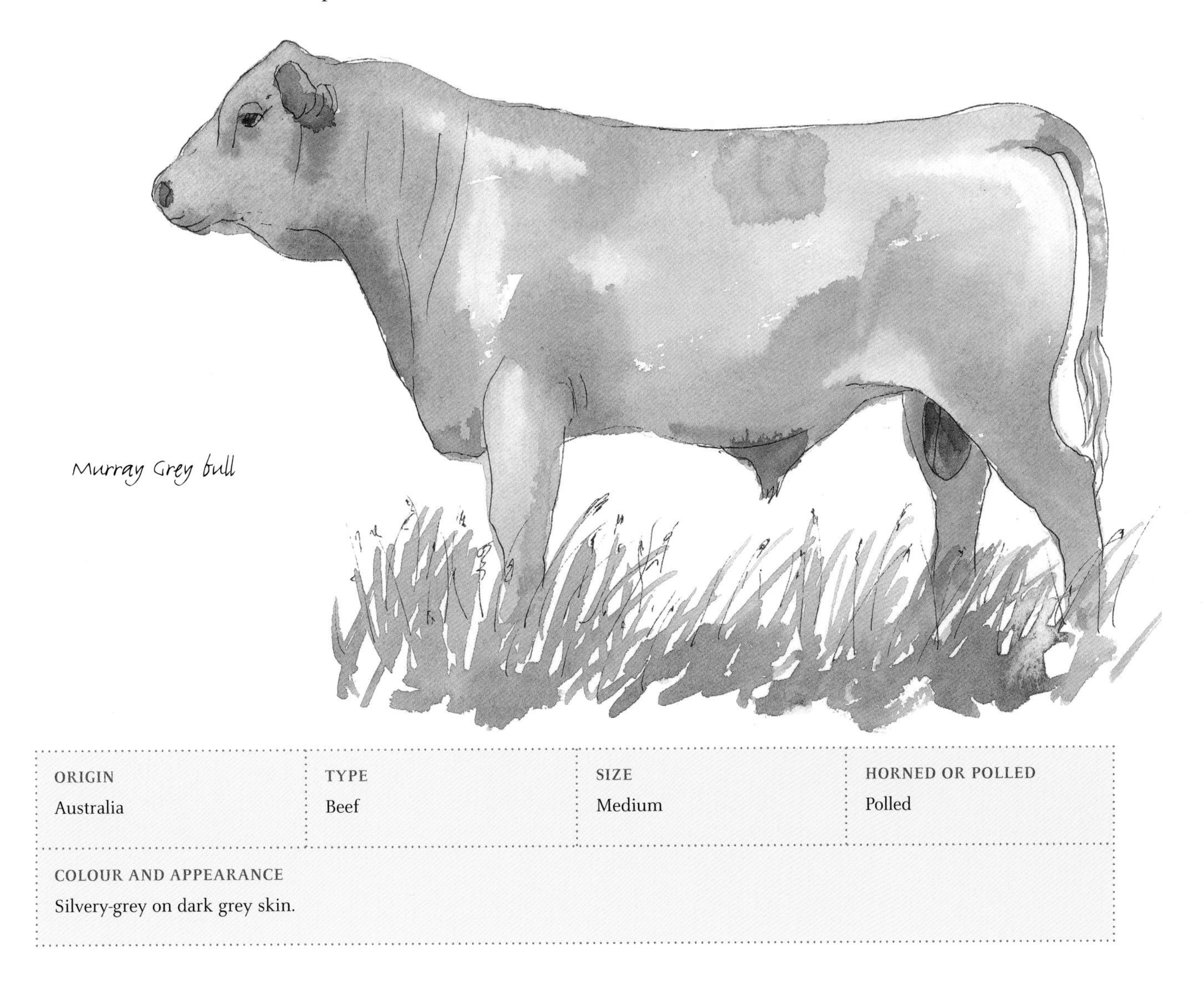

Murray Grey bull

ORIGIN	TYPE	SIZE	HORNED OR POLLED
Australia	Beef	Medium	Polled

COLOUR AND APPEARANCE
Silvery-grey on dark grey skin.

NORMANDE

The quintessential cow

Normande cow

According to the British Normande Cattle Society, the Normande is the third most popular milk breed in France. It is a true dual-purpose breed, producing copious amounts of milk that is highly prized in the cheese industry in Normandy, as well as tender, well-marbled meat. The first animals are said to have been brought to France by the Viking conquerors in the 9th and 10th centuries. They are now found all over the world, and in particular in North and South America.

The Normande's coat includes three colours, which are known as the three Bs: blanc (white), blond (fawn/red) and brindled (brown). These colours result in three types of coat: quail, white with scattered patches of colour; brindled, predominantly brown; and trouted, with a multitude of brown spots on the skin beneath the white hair. The animals have distinctive eye patches that protect their eyes from bright sunlight in the summer.

This is an adaptable breed that is able to live at high altitude and travel great distances over rough terrain. The females reach sexual maturity at an early age and calve easily. The breed shows remarkable docility – a trait carried by the bulls as well as the cows.

ORIGIN	TYPE	SIZE	HORNED OR POLLED
France	Beef and dairy	Medium to large	Small horns

COLOUR AND APPEARANCE
Black or red with white spots or brindling. Dark eye patch.

NORTH DEVON

A docile possessor of the thickest hide

Red Ruby, known as the North Devon to distinguish it from the South Devon, is one of oldest breeds in existence – in fact, its origins may well be prehistoric. This was originally a multi-purpose breed that was used as a draught animal as well as for its beef and milk. It has evolved through careful selection into a beef animal, although there is a Milking Devon strain that is unique to the USA. Devon cattle reached the USA in 1623 with the Pilgrims, and were probably the first pure-bred cattle to reach North America.

North Devon cow and calf

Devons have exceptionally docile temperaments and a natural resistance to disease. They do well on grass-based systems and thrive on poor land as conservation grazers. They are exceptionally tolerant of variations in climate, hot or cold, having the thickest hides of any cattle in the world, which also gives them resistance to parasites. Cows give birth to small, vigorous calves with little calving trouble, and for this reason Devon bulls are in high demand for cross-breeding of both beef and dairy animals. The cows make excellent mothers and their calves have one of the highest survival rates to weaning of any breed.

The first American corned beef was made from a Devon carcass.

ORIGIN	TYPE	SIZE	HORNED OR POLLED
UK	Beef	Medium	Medium horns

COLOUR AND APPEARANCE
Red

PARTHENAIS

A muscly producer of very lean beef

The Parthenais bears a passing resemblance to the Limousin, but has black points in common with many old breeds – this is a particularly old breed that has had a herd book since 1893. Parthenais were originally draught animals, working in the fields and pulling carts. Coming from the Deux-Sèvres region, and the town of Parthenay in western France in particular, they are well adapted to coping with most climates and all types of farming system, from intensive to pasture.

The Parthenais is double muscled but not excessively so, and despite its size calves easily due to its light bone structure and none of the double muscling that will develop later showing at birth. The calves are dark brown when born, but soon become the typical rusty-red with a black muzzle, eyes, ears and tip of the tail. This is a hardy breed with particularly strong legs and feet, making it suitable for almost any terrain. The beef is very lean and low in cholesterol – this is supposedly lower than it is in chicken – with high yields.

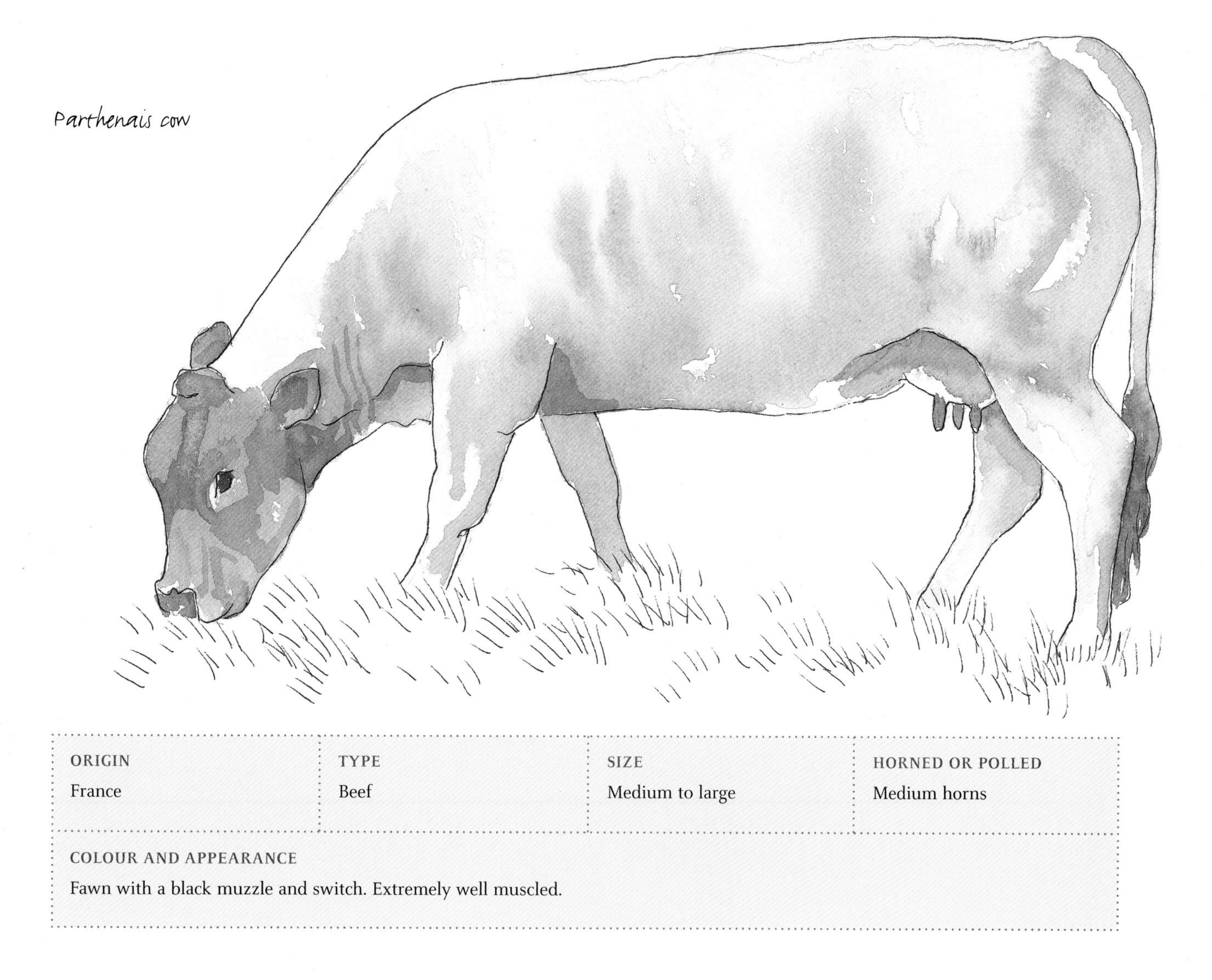

Parthenais cow

ORIGIN	TYPE	SIZE	HORNED OR POLLED
France	Beef	Medium to large	Medium horns

COLOUR AND APPEARANCE

Fawn with a black muzzle and switch. Extremely well muscled.

PIEDMONTESE

A double-muscled breed with a calm temperament

Piedmontese are recent arrivals on British shores, not having been imported until 1988, although they reached the US and Canada a few years before this. Native to the Piedmont area of northern Italy, they adjust easily to cold, damp winters and hot, dry summers, and have good foraging ability. This is a very ancient breed thought to be descended from the Aurochs as seen in prehistoric cave paintings.

Piedmontese cow and calf

Although originally a triple-purpose animal, the Piedmontese has been carefully selected for beef and now tends to be double muscled, similarly to the Belgian Blue. Piedmontese beef is particularly low in cholesterol, having just 48 mg per 100 g beef, compared with other beef (usually 73 mg cholesterol) and chicken (76 mg).

Piedmontese are exceptionally calm animals, calve well and have a strong maternal instinct. The cows are also long-lived and can go on calving for up to as much as 17 years of age.

ORIGIN	TYPE	SIZE	HORNED OR POLLED
Italy	Beef	Medium	Short horns

COLOUR AND APPEARANCE

Pale grey to light grey with black eyes, muzzle and switch. Calves are beige when first born.

PINZGAUER

A red-and-white breed from the mountains

As long ago as AD 500 there were red-and-white cattle grazing in the alpine pastures of Europe. They were bred selectively to withstand mountain conditions while producing adequate milk and good beef, and in ancient times they were also used as draught animals. The name Pinzgauer derives from the Pinzgau region of Austria and has been documented since the 1600s. Herd books dated 1700 exist, and records show that Pinzgauer cattle were exported to Romania, Czechoslovakia and Yugoslavia in the 1820s.

Pinzgauers are ideally suited to harsh conditions, whether very hot or cold, and have adapted to life in situations ranging from the icy conditions of Canada to hot and barren South Africa. Their almost mahogany-red and white coats have the 'lineback' pattern that refers to animals with a prominent white stripe along the back and white underbelly. Their coats are short and dense, and the pigmentation of the skin protects it from sunburn, particularly around the eyes, making eye disease virtually non-existent in this breed.

Females reach sexual maturity early and can continue to breed well into their teens, producing calves that gain weight readily. Due to their quiet temperament, they adapt easily to life in the feed yard, but also do well on pasture alone and reach market weight without the use of grain.

Pinzgauer cow and calf

ORIGIN	TYPE	SIZE	HORNED OR POLLED
Europe	Beef	Large	Medium horns or polled

COLOUR AND APPEARANCE
Deep red and white with white lineback.

RED POLL

An adaptable possessor of a quiet disposition

The counties of Suffolk and Norfolk are the traditional home of the Red Poll, which was originally developed as a dual-purpose breed. Red Polls are thought to have been the result of crossing the Norfolk Red with the Suffolk Dun, both of which are now extinct breeds. The Suffolk Dun was a naturally polled breed and this trait was passed on. The animals have the reputation for fattening easily on poor-quality grazing, and although not large in size produce a good-quality carcass as well as plenty of milk.

Red Poll cow and calf

As their name suggests, Red Polls are red – a deep, rich mahogany-red – and some animals may have a white tail switch and sometimes white areas around the udder. They are long-lived and adaptable creatures with quiet dispositions leading to ease of handling, and are ideal as house cows. Their milk is high in protein and has smaller fat globules than the milk of some other breeds, making it ideal for cheese making.

ORIGIN	TYPE	SIZE	HORNED OR POLLED
UK	Beef and dairy	Medium	Polled

COLOUR AND APPEARANCE
Dark mahogany-red, with occasionally a white tail switch.

RIGGIT GALLOWAY

A native hill breed with an ancient pedigree

According to the Riggit Galloway Cattle Society this breed of cattle is a well documented archaic strain of Galloway, easily identifiable by the white stripe running down their spine. The term 'riggit' is a Scottish vernacular reference to this stripe, and seems to be Scandanavian in origin. This coloration is also known as finch-back or line-back and also occurs in Gloucesters and Pinzgauers, among others. The main body colour can be black, blue / black, red, brown or dun. The white coloration may include a widening of the stripe to cover much of the back – particularly on the hindquarters – white under the keel of the animal, and white flashes amongst the solid colour.

The Riggit is a medium-sized animal with a good beef conformation. It is a native hill breed well suited to wintering out in harsh conditions thanks to its double coat, which has a well insulated undercoat and a longer outer coat to shed rain and snow. This breed is well liked by conservation grazing schemes as the animals are not too large and therefore don't over poach the ground, instead they churn it up just enough to encourage germination of plants. The beef is well marbled and tender and apparently 'tastes like beef used to taste'.

Although the breed very nearly died out at the beginning of the 20th century and is still extremely rare, the Riggit Galloway Cattle Society was formed in 2007 with the intention of consolidating the breed's future.

Riggit Galloway cow and calf

ORIGIN	TYPE	SIZE	HORNED OR POLLED
UK	Beef	Medium	Polled

COLOUR AND APPEARANCE
Black, blue / black, red, brown or dun with white stripe along the back and hind quarters.

ROMAGNOLA

A muscly beast ideal for a hot climate

The Romagnola is descended from the wild ox that moved down the Italian peninsula from the steppes in the 4th century, and early cattle brought by Aginulf with the Goth invasion. Several breeds developed from these primitive cattle, the Romagnola being one, the Chianina another. In ancient times the breed was used as a draught animal for ploughing, with beef as a secondary product.

With mechanization the breed changed from being used as a draught animal to one utilized primarily for beef production. The founding father of the modern-day animal is considered to be Medoro, a bull born in 1920 near Ravenna. He served at stud for 13 years and was responsible for changing the basic structure of the cattle towards a modern type, where muscling and stockiness were the main characteristics.

The Romagnola has black-pigmented skin, a pale grey coats and black points, and like other cattle with this colouring the calves are born a fawn colour and do not become grey until three months of age. Cows give birth to small calves, but because their milk is rich the calves grow fast. The dark pigmentation provides protection from the sun, making the breed ideal for hot climates. Among typical characteristics are the horns, which are lyre shaped in the cows and half-moon shaped in bulls.

Romagnola bull

ORIGIN	TYPE	SIZE	HORNED OR POLLED
Italy	Beef	Medium	Medium horns

COLOUR AND APPEARANCE

White becoming darker grey towards the head, with black points. Calves are fawn at birth and become grey at three months. Well muscled and stocky.

SALERS

A lively breed with ancient ancestors

The Salers, pronounced 'sa'lairs', is thought to be one of the oldest breeds of cattle in the world, with prehistoric cave paintings showing that a similar type of cattle existed 10,000 years ago. The breed originated in the Massif Central in France, in particular the Auvergne, where the milk is used in the region's well-known cheese, Bleu d'Auvergne. The first Salers to reach the UK were imported to Cumbria in 1984.

salers bull

Like most old breeds, this was originally a triple-purpose animal that survived well on the poor forage in the tough climate of its upland home. By the mid-19th century other breeds, such as Devons, were imported into the area by a M. Tyssandier d'Escous, who took it upon himself to improve the breed by careful selection. His contribution to the success of the breed is commemorated with a statue in the small town of Salers.

Salers are well suited to poor grazing and grow a good winter coat. They are particularly known for their strong legs and feet, with good black hooves that stand up to rough ground. Cows have strong maternal instincts, and give birth with ease to small but vigorous calves that grow fast. Salers have the reputation of having wild natures, but careful breeding for docility and sensible handling can overcome this issue.

salers cow and calf

ORIGIN	TYPE	SIZE	HORNED OR POLLED
France	Beef	Medium	Medium horns or polled

COLOUR AND APPEARANCE
Deep red, but a black form exists in the US.

SENEPOL

An island dweller that loves the heat

The island of St Croix, one of the US Virgin Islands, is the home of the Senepol. In the 1800s N'Dama cattle had made their way to the island via slave traders, and Henry C. Nelthropp of the Grenard Estate became one of the largest N'Dama breeders, with over 250 head. The cattle adapted well to the climate and poor vegetation, but were lacking in the milk-production department. Nelthropp's son Bromley decided to create a breed that would combine the traits needed for good production in the tropical environment, and bought a Red Poll bull from the neighbouring island of Trinidad. With careful selection for polled animals of a solid red colour and gentle disposition, the Senepol breed was established.

Due to the isolation of its island home the Senepol breed has remained pure, and only the progeny of the best females has gone back into the herd. In 1977 the first cattle were exported to the US, where they are used to produce beef on poor grazing in the hot and humid conditions that they are used to. The animals have very short hair of a solid red colour, athough this can be dark red through to a lighter ginger colouring. It is typical to see Senepol cattle grazing in the open in the heat of the day when other breeds would have sought shelter in the shade.

Senepol cow and calf

ORIGIN	TYPE	SIZE	HORNED OR POLLED
Virgin Islands	Beef	Medium	Polled

COLOUR AND APPEARANCE

Deep red to lighter ginger.

SHETLAND

A tough and hardy house cow

Not surprisingly, this is a tough and hardy breed that can cope well with a harsh climate and poor forage; it is even capable of eating seaweed when there is a lack of anything more nourishing.

This is an ideal smallholder's animal, and in fact it was traditionally a house or crofter's cow with all the attributes required. Shetlands are small and therefore easier to deal with, eat less and do less damage to the ground than larger breeds. They are exceptionally healthy and naturally docile – another important consideration for the smallholder. Cows give birth with ease to small but vigorous calves that grow fast.

Shetland bull

Like many similarly resilient breeds, the Shetland is being used in conservation grazing of lowland bog, coastal marshland and heathland, especially on wet sites where out wintering is necessary. Sadly, the Shetland is now on the Rare Breeds Survival Trust's 'at risk' list.

Shetland heifer

To laugh is human
but to moo is bovine.
Author unknown

ORIGIN	TYPE	SIZE	HORNED OR POLLED
UK	Beef and dairy	Small to medium	Short horns

COLOUR AND APPEARANCE
Black and white, but occasionally red and white.

SIMMENTAL

The most adaptable of large cattle

This old breed originated in the Simmen Valley in western Switzerland where, like many ancient breeds, it was originally a dual-purpose animal. Occurring in many different colour variations, the typical coloration comprises a dun or reddish body and a pale face. The white face is characteristic and usually passed to cross-bred offspring. In the US the most common colour is black. The breed is horned or naturally polled. Simmentals are particularly large animals, only the Brahman being larger, and the breed is now the most numerous in Europe.

Simmentals are adaptable creatures and are now found all over the world in differing climates. In many countries they are used as dual-purpose beef and dairy animals. In the UK they have been developed solely as beef animals, and are used by both large farming operations and smallholders.

simmental cow and calf

simmental cow

A bull in a china shop: to be very clumsy in a delicate situation.

ORIGIN	TYPE	SIZE	HORNED OR POLLED
Switzerland	Beef and dairy	Large	Medium horns or polled

COLOUR AND APPEARANCE
Varies from red and white, through yellowish-tan, to black.

SOUTH DEVON

Gentle giants of the bovine world

Large red cattle are thought to have reached British shores with the Norman invasion, and it is from these that the South Devon originated in the South Hams of Devon. It is thought that South Devons reached North America on the *Mayflower*. Originally used as draught animals as well as for their beef and milk, since the Second World War South Devons have become solely beef animals.

These very adaptable creatures are now found on every continent and are the largest of the native British breeds, producing leaner carcasses than many breeds. They are particularly prized for their intramuscular fat, known as marbling, which adds to the taste and tenderness of the beef.

Sometimes known as the Orange Elephant or Gentle Giant, this is a docile breed. The cows are early to mature and long-lived, can produce a calf a year for as long as 15 years and have strong maternal instincts.

South Devon cow and calf

To take the bull by the horns: to do something difficult in a determined and confident way.

ORIGIN	TYPE	SIZE	HORNED OR POLLED
UK	Beef	Large	Horned and polled

COLOUR AND APPEARANCE
Curly coat of a light red colour, and a pink muzzle. Stocky.

SUSSEX

An early-maturing breed from the Weald of Sussex

There have been Sussex Cattle on the Weald of Sussex and Kent for many centuries, and like many ancient breeds they were originally used as draught animals. They were also used for hauling timber and ploughing, and the Sussex Cattle Society reports the case of 'an ancient lady of good quality going to church in a country village, drawn in her coach by six oxen'.

Sussex are medium-sized animals that mature early, a fact that allows them to compete with larger, slower maturing continental breeds, particularly where the beef is produced off grass pasture. They are not fussy eaters and will make the most of any quality of grazing. This is a hardy breed with a thick winter coat that develops into a beautiful shiny summer coat, allowing it to adapt to heat but also enabling the animals to be out wintered even in exposed areas. The cows are fertile, calve easily and produce plenty of milk. The beef is well marbled, with a fine texture.

Sussex cow

Sussex cow

ORIGIN	TYPE	SIZE	HORNED OR POLLED
UK	Beef	Medium	Medium horns or occasionally polled

COLOUR AND APPEARANCE
Dark red with white tail switch.

TEXAS LONGHORN

The symbol of Texas with the longest horns

The Texas Longhorn is a direct descendant of the Spanish cattle brought to the Americas by Christopher Columbus in 1493. These cattle were tough and well suited to the sparse forage, harsh climate and vast distances of southern and central America. They could survive without water longer than most and were strong enough to be driven immense distances.

By the time of the American Civil War, it was thought that there were as many as five million Texas Longhorns wandering the open range. However, by the 1930s the open range began to be fenced and ranchers chose more commercial breeds from Europe to farm. The Texas Longhorn population crashed until the breed was almost extinct. In 1964 the remaining breeders got together and formed a breed association to try and promote the breed and maintain its link with American history.

Longhorns are known for their lean meat and long lives. They need less supplemental feed than many breeds due to their foraging ability. They are naturally immune to many diseases and parasites, and thrive in all types of climate. There is even value in their magnificent horns, which are used to adorn front doors in the south-western states, and can extend to an amazing 2 m or more (7 ft) from base to tip. Add to this the fact that they are intelligent, docile and handsome beasts, and it is hard to imagine why they became endangered in the first place.

Texas Longhorn steer

ORIGIN	TYPE	SIZE	HORNED OR POLLED
USA	Beef	Medium	Very long horns

COLOUR AND APPEARANCE

Wide variety of colours – no two are alike. Striking horns.

WATER BUFFALO

A placid producer of very rich milk

These placid creatures are still used as draft animals in Asia as well as providing milk, meat, hide, horn and fuel. They have been domesticated for 5,000 years and today still provide more than 5 per cent of the world's milk supply. Their lean and flavoursome beef contains less than one-fourth the fat of normal beef, but they are chiefly known for the production of mozzarella as their milk is exceedingly rich having less water and more fat, lactose and protein than cows' milk.

Water Buffalo are classed as normal cattle and can be contained by suitable fences. They thrive on poor grazing and consume a fraction of the feed required to raise other cattle. Calving problems are almost non-existent and the cows have good mothering instincts. They are intelligent and long-lived, frequently still calving into their twenties.

This is a calm and easy-going breed that can easily be trained as a draft animal and will also even allow itself to be ridden (if that is your desire!). As you might imagine they also love water and are never happier than when wallowing in mud creating an insect- and sunburn-proof coat.

There comes a time in the affairs of man when he must take the bull by the tail and face the situation.
W. C. Fields

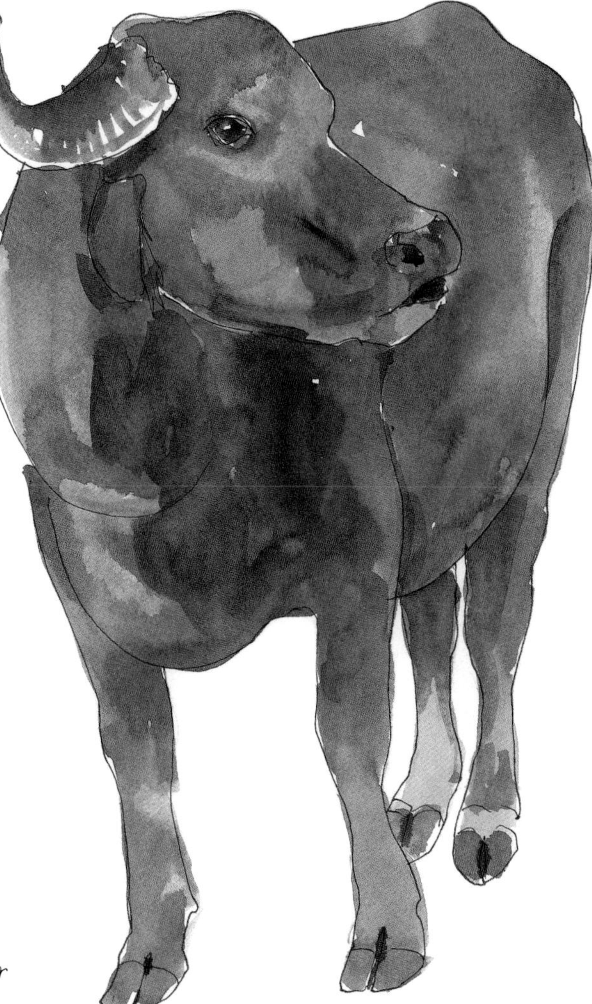

Water Buffalo heifer

ORIGIN	TYPE	SIZE	HORNED OR POLLED
Asia	Beef and dairy	Large	Large horns

COLOUR AND APPEARANCE
Dark grey when dry but dark brown to black when wet and almost hairless.

WELSH BLACK

Is this the black gold from the Welsh hills?

The origins of the Welsh Black are lost in the mists of time, but it is thought that its ancestors were around in Roman times and that their forebears may have come from the Iberian Peninsula. Black cattle have lived in Wales for well over 1,000 years, and according to the Welsh Black Cattle Society were once used as currency, giving rise to the description 'the black gold from the Welsh hills'.

Until the 1970s the breed was thought of as dual purpose, although there were two distinct strains: a stocky beef type from North Wales and a more dairy-like animal in South Wales. The modern breed combines the best of both strains and has the reputation of thriving on the poor forage of the hills, growing a good, thick coat in winter that is resistant to rain and snow.

Welsh Blacks are docile cattle that make excellent mothers and live to an old age. They are mainly black, although occasionally they have rusty to red patches. Their horns are white with black tips, and there is also a naturally polled strain.

Welsh Black cow and calves

ORIGIN	TYPE	SIZE	HORNED OR POLLED
UK	Beef	Medium	Medium horns; polled strain also exists

COLOUR AND APPEARANCE
Mainly black but occasionally red; some white on the underbelly may occur.

WHITE PARK

An ancient breed with aristocratic looks

This ancient breed has been found in Britain for more than 2,000 years. According to the White Park Cattle Society, it is closely descended from Britain's original wild white cattle, which were enclosed in parks by the nobility during the Middle Ages. During the Second World War the White Park breed was considered by the government to be sufficiently important as a part of the British heritage for a small unit to be shipped to the USA for safe keeping. Despite this, the breed almost reached extinction until in 1973 the Rare Breeds Survival Trust was formed and chose the White Park breed as its logo. From as few as 60 remaining animals, numbers have grown until there are now well over 700 adult breeding cows and the future looks safe for the breed.

These are good-looking animals, with their white coats and smart black points on the muzzle and ears, and around the eyes. They should not be confused with the British White (see page 46) – although they look similar they are genetically separate breeds. White Park cattle have elegant horns that are also black tipped. Cows have the reputation for easy calving, plenteous milk production and high fertility, and bulls are used as crossing sires to pass on these traits. They are also exceptionally long lived, some breeding until they are 20 years of age.

White Park cow and calves

ORIGIN	TYPE	SIZE	HORNED OR POLLED
UK	Beef	Large	Large horns

COLOUR AND APPEARANCE
White with black points.

WHITEBRED SHORTHORN

The hardiest of breeds with the thickest of coats

The exact origin of the Whitebred Shorthorn is lost in the mists of time, but it was once known as the Cumberland White. The Whitebred Shorthorn is a completely different breed from the Dairy and Beef Shorthorns. When crossed with the Galloway cow, a recognized crossbreed known as the Blue Grey is achieved. Bulls are also crossed with the Highland, producing a Cross Highlander that is well suited to the full range of British climatic conditions.

The Whitebred Shorthorn has a creamy-white coat with two layers – a soft outer layer and a mossy undercoat – making it impervious to the harshest of weather and capable of wintering outside where the land is suitable. It is long-lived, producing calves well into its teens, and exceptionally docile, making management easy.

The Whitebred is also used in conservation grazing, particularly for upland management, because it requires little supplementary feeding. According to the Rare Breeds Survival Trust its white colouring is an advantage in that it make the cattle relatively easy to spot.

Whitbred shorthorn bull and heifer

ORIGIN	TYPE	SIZE	HORNED OR POLLED
UK	Beef	Medium	Horned

COLOUR AND APPEARANCE
Creamy-white.

How to milk a cow

Cows are creatures of habit, so always milk your cow at the same time each day if possible, preferably 12 hours apart. Keep her happy by feeding her at the same time.

If you have never milked a cow and have one that has never been milked, it is a good idea to first have a go on a cow that is used to being milked to get the feeling.

Cleanliness is important: wash your hands and the udder before you start.

Sitting on a stool, lean against the right-hand side of the cow.

Place the bucket underneath, between your knees if possible – remembering that cows are adept at kicking you or the bucket over if they so wish. Talk quietly to the cow to reassure her. Some cows 'let down' with ease, but others may take a little coaxing to begin with.

Milk the cow,
but do not pull off the udder.
Greek Proverb

Starting with the front two teats, and using one hand per teat, close your thumbs and forefingers around the tops of the teats (this prevents the milk in the teats going back up into the udder). Using your fingers sequentially, gently squeeze out the milk, directing the stream into the bucket. Use one hand at a time and try and get a rhythm going – there is no need to pull. Release your thumbs and forefingers and the teats will refill with milk.

It takes about 350 squirts for each gallon of milk from a cow.

When the front teats no longer produce any milk, move on to the rear two. It is important to drain each quarter with care or the cow will gradually dry up.

A cow can produce 45 litres (10 gal) or so of milk at a time and your fingers will get tired at first – you may even suffer from cramp, but this will ease with practice. A good milker with an easy cow can milk about 4 litres (1 gal) in five minutes.

Dairy products

Cream

milk churn

You need to skim the cream off your milk if you want to make butter or clotted cream. As in the case of milking, everything must be scrupulously clean. You need a large, wide, shallow pan, preferably not made of plastic – traditional ones are made of pottery. Leave the freshly milked milk in a cool place for 24 hours so that the cream rises to the top. Use a cream scoop (or saucer if you do not have one) to skim off the cream. As a rule you will require 4–5 litres (8–10 pints) of milk to produce ½ litre (1 pint) of cream.

Clotted cream

Clotted cream keeps longer than fresh cream and can also be used to make butter. Leave the milk in a shallow, heatproof pan in a cool place for 24 hours. Then very carefully heat to 65–75°C (150–170°F). A ring will form around the edge of the pan and the cream will split away. Leave the pan to cool for at least 12 hours, then skim off the cream.

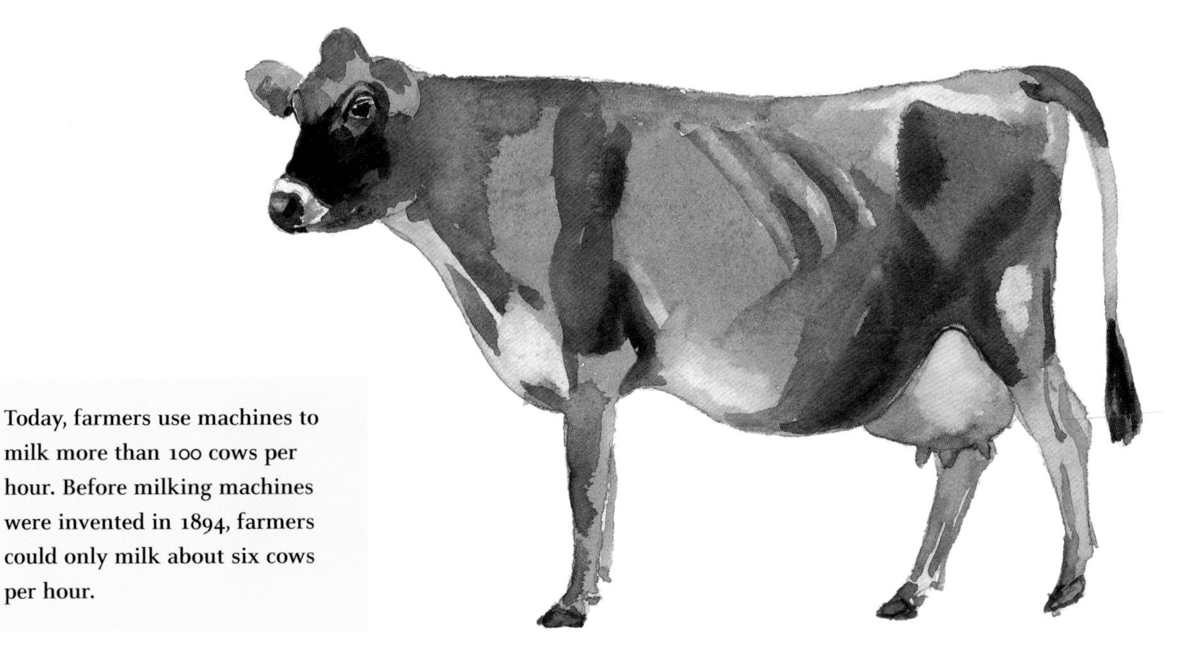

Today, farmers use machines to milk more than 100 cows per hour. Before milking machines were invented in 1894, farmers could only milk about six cows per hour.

Butter

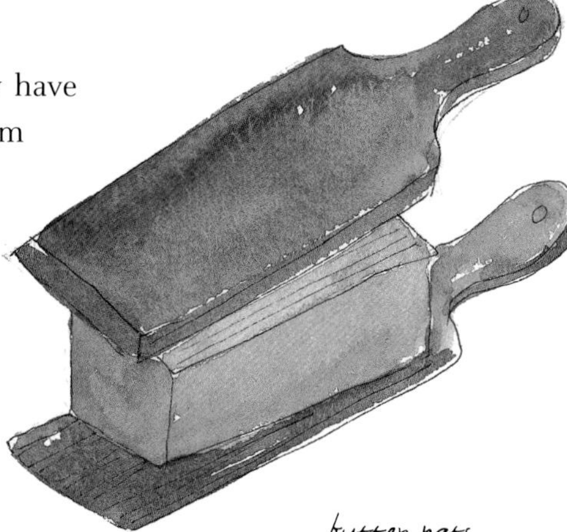
butter pats

Making butter is remarkably simple – you may have made some by mistake by over whipping cream and basically that is exactly what you have ended up with. You can use an commercial electric churn if you have a great deal of cream. A less-expensive option is a glass jar with a handle that turns a paddle inside, or you can simply use a kitchen food processor.

1. Collect the cream as described above.
2. Leave the cream to 'ripen' in a warm room at 2°C (68°F) for 12 hours.
3. Place the cream in the churn or food processor with the beating blade attached, and beat it.
4. The cream will gradually thicken until it becomes quite stiff.
5. The cream will suddenly darken in colour and break into little bits, and the thin buttermilk will appear. Slow the food processor as this happens.
6. Drain off the buttermilk (you can use it in baking or drink it).
7. Add some clean cold water to the butter in the churn and operate at low speed for a minute.
8. Drain off the water and repeat until the water remains completely clear.
9. Squeeze out all the water from the butter using wooden butter paddles (or wooden spoons or even your hands).
10. Add a little salt to taste and to help preserve the butter, mixing it in well until the butter is smooth.

glass butter churn

The natural yellow colour of butter comes mainly from beta-carotene found in the grass the cows graze on.

Yoghurt

This is one of the easiest things that you can make with your own milk. Buy a small pot of plain yoghurt (ignore ones that say 'live' – all yoghurt is live).

Milk is best for cooling your mouth after eating spicy food. Milk products contain casein, a protein that cleanses burning taste buds.

1. Place 500 ml (1 pint) milk in a pan.
2. Heat carefully until the milk begins to steam and you can just see bubbles forming around the edge of the pan.
3. Pour into a bowl and allow to cool to 46°C (115°F), or until you can keep a finger in the milk for 20 seconds without pain.
4. Whisk in 2 tablespoons plain yoghurt and cover the bowl with a clean dishcloth, lid or clingfilm.
5. Place the bowl somewhere warm such as the back of an Aga, in an airing cupboard or in the absence of either in a thermos flask.
6. Leave for at least six hours or overnight until the yoghurt has set.

If you like your yoghurt thicker, adding a little dried milk powder to the milk before you heat it will do the job.

You can make a sort of light cheese by carefully placing the yoghurt in a jelly bag and leaving it to strain. You will be left with a thicker substance that can be rolled into balls and used a bit like mozzarella – add salt, pepper and any herb you fancy.

Curd cheese

For this you need to curdle the milk. You can do this by adding lemon juice, but rennet will produce more successful results.

1. Place 2 litres (4 pints) milk in a saucepan and heat carefully to blood temperature.
2. Remove from the heat and add 2 teaspoons rennet (or the juice of 2 lemons). Let the milk stand for 15 minutes or so until it has separated.
3. Carefully collect up the curds with a slotted spoon and put them in a jelly bag, leaving it to drip for three or four hours.
4. Remove the curd from the jelly bag and put it in a container. Store in a fridge.

Chopped chives go particularly well with this form of cheese.

Finishing

Finishing is an art in itself and means bringing the animal to the best possible condition for killing. There should be not too much fat, and not too little, with the right amount of muscle in all the right places. At present in the UK all meat animals have to be slaughtered before they are 30 months of age (there are special rules for slower growing animals such as Galloways). Introducing and carefully controlling the use of concentrates (see page XXX of Feeding) is the key and depends entirely on your breed and method of rearing.

Cuts of beef

The carcass will be made up of the following:

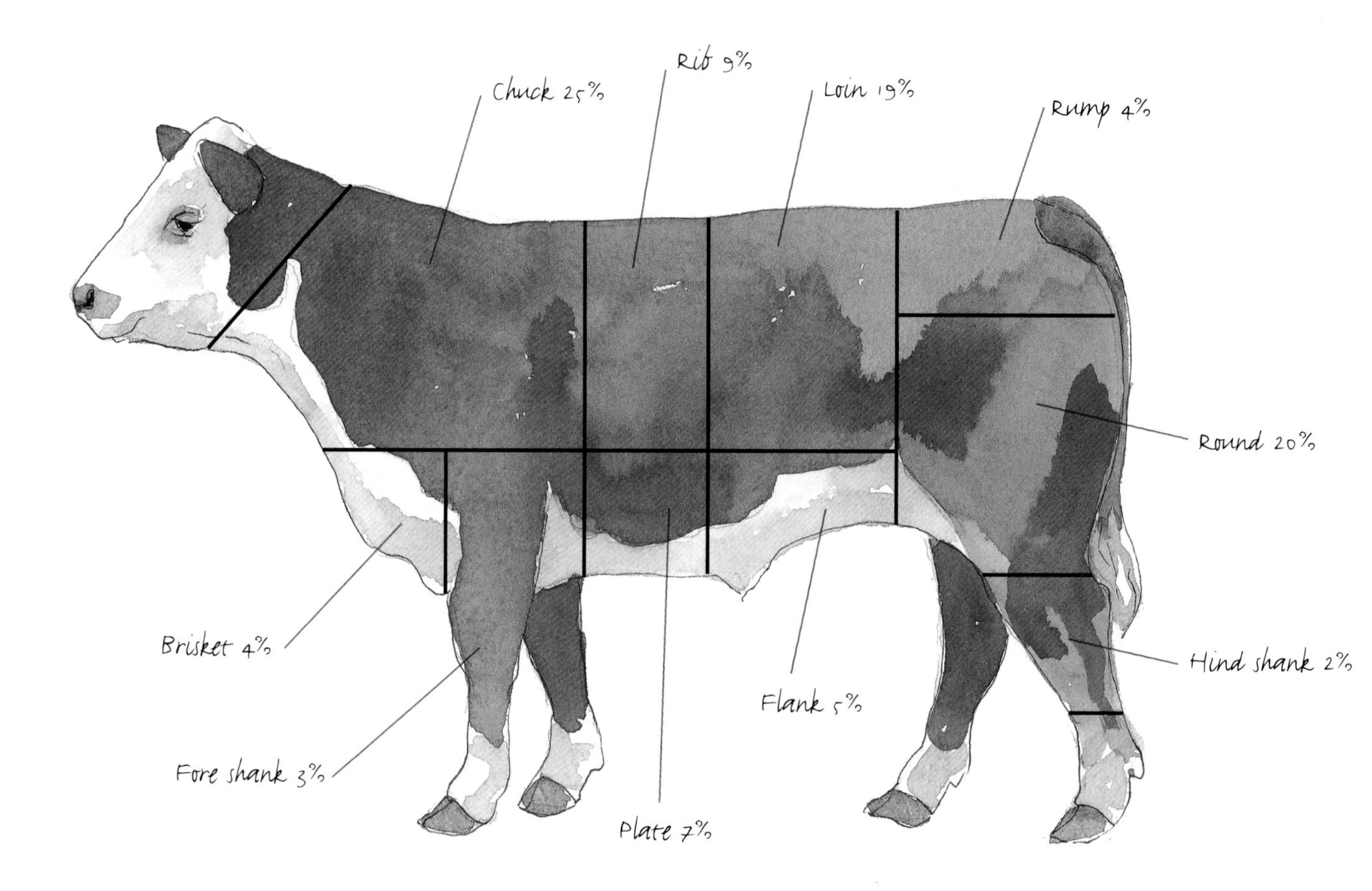

Cow-horn bugle

A cow horn was possibly man's first musical instrument, but there are many other useful and beautiful uses for it. Horn can be turned into spoons, buttons, drinking vessels, combs, bracelets and other jewellery, shoehorns and even bowls. In order to get back the horns of your animals from an abattoir, in the UK you need a Request Form AB117 and will only be allowed to obtain the horns unattached to the skull.

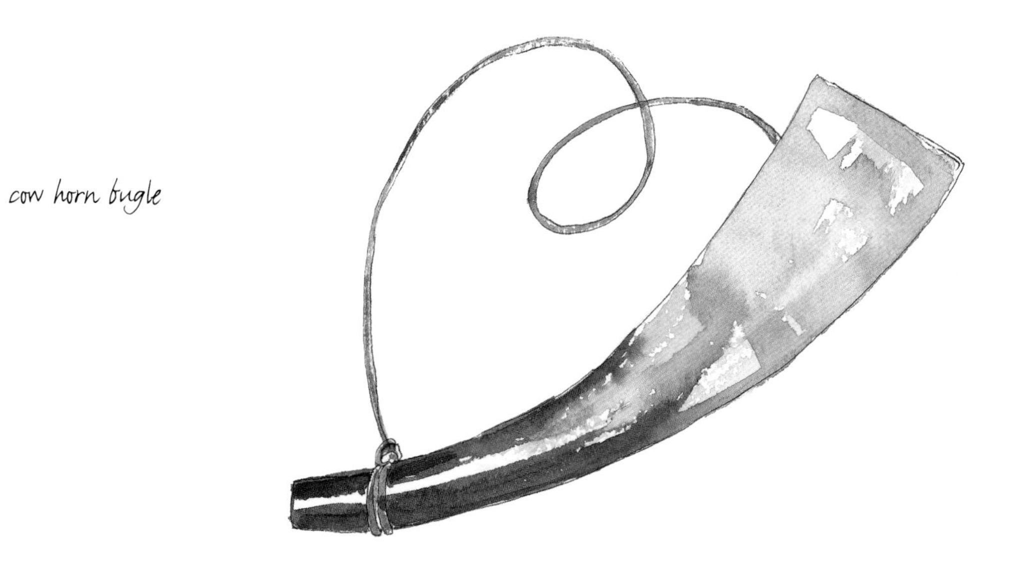

Imagine how satisfying it would be to be able to call your own cattle by blowing on a horn you have made yourself. Here is how to make one:

How to make a cow horn

First you must clean out the soft core of the horn. This is a smelly job and there are various ways of doing it.

Soak the horn in boiling water until it softens a little.

Dry out the horn by keeping it in a hot place such as the back of an Aga, beside a boiler or on a hot surface outside in the summer. As the horn dries out, the inner and outer parts separate, making it easier to remove the insides.

Leave the horn soaking in cold water for a few months. This method can create an unpleasant smell unless the water is regularly changed.

Whichever method you use, you will then have to scrape out the inside of the horn using any tool that will fit inside it.

Cut off about 2.5 cm (1 in) from the pointed end of the horn (you can use this to make a toggle) for the mouthpiece and about 1¼ cm (½ inch) from the wide end to prevent splitting.

Sand and polish the horn to your heart's desire and finish by buffing with a little oil and a soft cloth.

Blow the horn with pursed lips. The smaller the horn, the harder you will have to blow and the higher will be the note – horn blowing is an acquired art.

some of the other many things that can be made from cow horn

Tanning a hide

When your animal goes to the abattoir, the hide is taken into account in the slaughter charge and will be sent off to a tannery. Most abattoirs will not guarantee that you will get your own cow's hide back from the tannery if requested, so it may be that you would like to tan your own. Hides are classed as by-products, and in the UK you will need a Request Form AB117 to take to the abattoir.

Tanning is a method of treating skins to produce leather and make the skin more durable and less susceptible to decomposition. It is a lengthy and smelly business and traditionally tannin, an acidic chemical, was used. The word comes from the Latin *tannum* (oak bark), which is used in some methods. Other methods employ urine and even brains – rather delightfully, each animal, regardless of size, has the right amount of brain to tan its own skin.
Here is a method of tanning hides using less smelly and disgusting materials. Wear rubber gloves throughout and dispose of excess liquid with care:

1. As soon as the hide is off the animal, scrape away as much flesh as possible with a blunt knife (a sharp knife might puncture the hide).
2. Lay the hide out flat, flesh side up, and cover it with a good quantity of salt – if you lay it over a table and tip up the table by placing bricks under the legs on one side a good deal of liquid will drain.
3. Leave for two weeks.
4. Fill a tank with water. Submerge the hide for two hours. Scrape again.
5. Re-submerge for three hours or until soft.
6. Place 30 ml (1 oz) borax and 30 ml (1 oz) household detergent per 4.5 litres (1 gal) of water in the tub with the hide, then stir about, keeping the hide in for five minutes.
7. Rinse the hide with clean, warm water, then squeeze and allow to drip dry.
8. Mix 13.5 litres (3 gal) water with 1.6 kg (3½ lb) soda ash (soda crystals) and 2.7 kg (6 lb) salt. Stir until dissolved.
9. Separately mix 41 litres (9 gal) water with 680 g (1.5 lb) alum, then stir to dissolve and mix together with the prepared soda ash liquid.
10. Pour 18 litres (4 gal) of the solution into a large tank along with 145 litres (32 gal) clean water.
11. Submerge the hide for three days, stirring at least six times a day.
12. Remove the hide from the solution. Add half the remaining solution to the liquid, then return the hide to the tank and repeat the last step.
13. Repeat this step once more adding in the final third of the solution.
14. Remove the hide and empty the tank. Refill the tank with 91 litres (20 gal) clean water. Stir in 450 g (1 lb) borax. Soak overnight, remove the hide and allow it to drip dry.
15. Finally, after it has dried, replace the moisture in the tanned hide by rubbing it with a cloth soaked in neatsfoot oil.

Useful websites

www.bcms.gov.uk
The British Cattle Movement Service (BCMS) is the organisation responsible for maintaining a database of all bovine animals in the UK (except Northern Ireland).

www.defra.gov.uk and **www.gov.uk/defra**
The website of the United Kingdom's Department for Environment, Food & Rural Affairs (Defra).

www.thecattlesite.com
Latest cattle-industry news with farming features and information on managing cattle health.

www.rbst.org.uk
The website of the Rare Breeds Survival Trust.

www.soilassociation.org
The Soil Association, a membership charity campaigning for planet-friendly organic food and farming.

www.usda.gov
The website of the United States Department of Agriculture (USDA).

Acknowledgements

Aberdeen Angus: Angus Stovold
Aubrac: Deerpark Farm Services, Tipperary
Bazadaise: British Bazadaise Cattle Society
British White: Mary Dancey of Sheepdrove Organic Farm. Sarah Cook, Alcroft Livestock
Brown Swiss: Brown Swiss Cattle Society
Chillingham Wild Cattle: Chillingham Wild Cattle Association, Alnwick, Northumberland
Dairy Shorthorn: Jessica Miller
Friesian: Don Griffin; Mary Mead
Gascon: Gascon Cattle Society
Gelbevieh: Leigh Needham
Hereford: West Lavington Farms, Devizes, Wiltshire
Highland: Barbara R. Jones
Irish Moiled: Gillian Steele
Jersey: Pierrepont Farm, Frensham, Surrey
Kerry: Matthew English-Hayden
Limousin: Aled Edwards
Lincoln Red: Padworth Park Herd; Fenlady Herd
Longhorn: Gate Street Barn, Bramley, Surrey
Luing: Jamie Blackett
Maine Anjou: Eva Greer
Marchigiana: Associazione Nazionale Allevatori Bovini Italiani da Carne
Meuse Rhine Issel: Plawhatch Farm, East Sussex
Murray Grey: Brian Moody
Normande: Ann Pettigrew
North Devon: Jim Dufosee; Emma Bibby Jones; Catherine Broomfield
Piedmontese: Patrea L. Pabst
Red Poll: Maggie Edwards
Romagnola: Romagnola Beef Genetics
Salers: Laetitia Paris
Shetland: Shetland Cattle Breeders Association
Simmental: Morag Smith; Penny Lally
South Devon: Geoff Bryant; Margie and Mark Rushbrooke
Water Buffalo: The Corpe family, West Country Water Buffalo
Welsh Black: Janet Baxter
Whitebred Shorthorn: Whitebred Shorthorn Association

Index